I0001108

John Zephaniah Holwell

Dissertations on the Origin, Nature, and Pursuits of Intelligent Beings...

John Zephaniah Holwell

Dissertations on the Origin, Nature, and Pursuits of Intelligent Beings...

ISBN/EAN: 9783337025434

Printed in Europe, USA, Canada, Australia, Japan

Cover: Foto ©berggeist007 / pixelio.de

DISSERTATIONS

ON THE

ORIGIN, NATURE, AND PURSUITS,

OF

INTELLIGENT BEINGS,

AND ON

DIVINE PROVIDENCE,

RELIGION, AND RELIGIOUS WORSHIP.

In the Courſe of which,

The HONOUR and DIGNITY of

THE SUPREME BEING

IS VINDICATED

From the abſurd, if not impious Suppoſition, that by a *particular* or *partial Providence* HE interferes, influences, and directs, the Thoughts and Determinations of Individuals; and the political Government, Changes, and Events, of States and Kingdoms.

TO WHICH IS ADDED,

A neceſſary and moſt equitable Suggeſtion and Plan for the RELIEF of the PRESENT EXIGENCIES of the STATE, the BURDENS of the PEOPLE, and a more honourable Mode for SUPPORTING THE CLERGY.

Alſo, An eſſential Sketch for a more rational Form of Worſhip, and a NEW LITURGY.

By *J. Z. HOLWELL*, F. R. S.

Moſt humbly inſcribed, with all Duty, Loyalty, and Reverence, to the KING, (as Supreme Head of the Church)
And the Legiſlature of Great-Britain and Ireland.

BATH, printed by R. Cruttwell, for the AUTHOR;
And ſold by L. BULL, on the Lower Walks; T. CADELL, in the Strand; and C. DILLY, Poultry, London. 1786.

PREVIOUS to our entering on the diſcuſſion of our various ſubjects, it will be neceſſary, by way of introduction, to beſtow a few ſections on the relative nature and ſtate of man, wherein we ſhall not ſcruple, occaſionally, to *ſteal from ourſelf.* It is alſo neceſſary, that we account for the variations in ſentiment which will frequently occur in the courſe of theſe meditations, when contraſted with earlier productions ſubmitted to the public eye: To this our apology will be ſhort:—increaſe of years, experience, obſervation, and (we hope) juſt reflection, have produced theſe variations.

Sect. I.

AN inquisitive and philosophic genius will ask, why, and to what purpose, GOD created intelligent beings? We answer in the words and sentiments of the *most ancient Scripture*; at least, as far as our imperfect records tell.

§ 2.

" THE ETERNAL ONE, absorbed in the
" contemplation of his own existence, in
" the fulness of time, resolved to partici-
" pate his glory, and essence, with beings
" capable of feeling and sharing his beati-
" tude, and of administering to his glory.
" These beings then, were not; The Eternal
" One willed, and they were; He formed
" them in part of his own essence, capable
" of perfection, but with the powers of
" imperfection, both dependant on their

" voluntary election." And thus ſtands accounted for, the cauſes of the creation of the prime intelligent beings, or angelic bands.

§ 3.

HERE, the Supreme Being is repreſented as an abſolute, good, and powerful Sovereign, without ſubjects; which, in fact, is being no Sovereign at all. The cauſes aſſigned above, for this new creation, are ingenious and ſublime! ſome may think them ideal; but the *fact* remains beyond diſpute, intelligent beings do exiſt; the query is, can the cauſes for their creation be better accounted for?

§ 4.

HENCE it appears, that the powers of perfection and imperfection, or, in other words, the powers of doing good and evil, were coeval, and ſtrikingly marks the *free*

agency

agency or *independence* of the *will*, as the birthright of all intelligent beings.

§ 5.

THE Scripture before alluded to has the concurring teſtimony of all antiquity, and ſupport of every ſubſequent ſyſtem of theology reſpecting, not only the creation of angels, but alſo to their deviation from rectitude, and diſobedience to certain laws and injunctions they were ſubjected to at the period of their creation. It is generally ſuppoſed, that the number created was immenſe! but that only one-third of them rebelled. It would be derogatory to the Omnipotence of the Deity, to underſtand literally the phraſe, "And there was war in "heaven;" all that is here ſuggeſted, is, the rebels were ſubdued, tried, judged, condemned, and ſentenced to ſuffer certain puniſhments and degradations, for a *certain ſpace of time*, in a due proportion to the culpa-

culpability of the individual: for as they were, in part, emanations of their Creator's eſſence, he could not, or rather would not, annihilate them; annihilation precluding puniſhment.

§ 6.

THIS revolution in the celeſtial regions gave riſe to a ſecond *material creation*, flowing from the mercy and benevolence of the Supreme;—part of the ſentence of the delinquent angels was, baniſhment from his preſence, and expulſion from their ſeats and rank; and conſequently, they were plunged into an abyſs, dark, dreary, and horrid to imagination, as more particularly depicted in the ancient ſcripture before cited.

§ 7.

THE delinquents had remained in the abyſs for a *ſpace*, when the Creator relented, and graciouſly reſolved to mitigate their ſuf-

ferings,

ferings, and put it in their power to regain their loſt ſeats; in order to which, he formed the Planetary Univerſe for their reſidence, and releaſed them from the horrible abyſs.

§ 8.

The Eternal One formed his new material creation on ſuch occult principles and qualities, as it ſhould exiſt only *during the ſpace* allotted for the puniſhment and *probation* of the fallen ſpirits; by the laws of gravity, attraction, repulſion, &c. and by the *perpetual* action and re-action of *cauſes* to *effects* in all its parts; ſubject to ſuch changes as He, in his infinite wiſdom, ſhould meditate, and impreſs upon it. It was doubtleſs, in its original ſtate, glorious, lovely, and harmonious! worthy the Great Architect, ſuited for every rational enjoyment; befitting, and not unworthy the dignified beings doomed to inhabit it, although in a degraded ſtate, compared with the high rank they bore in the

preſence

presence of their Creator. Our conceptions, our judgment, should be limited to observations only, on the small portion of creation whose surface we are destined to occupy; all our conjectures, from science and analogy, on all beyond it, is uncertain, vain, futile, and unprofitable; and diverts our attention from the real end and purpose of the Deity's placing us here in this temporary state of probation. It has been for some time evident to the reader, that our chain of reasoning is founded on the presumption and full conviction, that the *souls* or *spirits*, animating every mortal organised form, are the identical apostate angels; but should any stumble at this pleasing, flattering, comfortable hypothesis, they are at liberty to reject it; as our essential arguments are equally applicable to *all*, considered as rational beings *only*. Yet we must not quit our subject, without laying before our readers those sentiments which influenced and determined us to this belief.

If

If we are not the apoſtate ſpirits, what are we? Whence came we? Why, and to what end and rational purpoſe, are we ſtationed here in an obvious ſtate of puniſhment and degradation? Why all this profuſion of wiſdom and contrivance that ſurround us, in the wonderful conſtruction, both internal and external, of the expanded univerſe, and of our own temporary mortal priſons? And laſtly, why this perpetual diſſolution and renovation of all theſe mortal organiſed and animated forms? Theſe amazing and ſtupendous operations of the Deity, can be no ways ſatisfactorily accounted for, but from this hypotheſis; and that we concluſively are the apoſtate ſpirits, ſuffering here for our pre-exiſtent lapſe. All concur in the belief, that the ſoul or ſpirit exiſts after the diſſolution of the body, ſo aſſuredly did it exiſt before its union with it, otherwiſe we muſt ſuppoſe the Deity is every inſtant employed in creating new ſouls;—a tenet, we think, the

moſt

moſt greedy and ravenous faith cannot ſwallow. But if we totally give up this hypotheſis, as merely ideal, yet none will diſpute, that man and brute are *intelligent beings* animating corrupt and mortal forms, whether they are the apoſtate ſpirits or not, (which by the bye is ſuppoſing we were all created at a venture) yet to theſe, as ſimply ſuch, our reaſoning and deductions are equally applicable, as above hinted.

§ 9.

NOR is the creation and formation of the various and innumerable mortal forms, allotted for the more immediate impriſonment of the offending ſpirits, a work leſs wonderful and ſtupendous! although marked and ſubjected to offices of the loweſt ſtate of degradation, deeply wounding to that fatal pride and ambition which firſt inſtigated their lapſe or fall.

§ 10.

WONDER riſes upon wonder, when we meditate on the incomprehenſible mode of animation given to theſe material forms, and the intelligent ſpirit's union with them in a kind of mutual dependance one on the other, and yet in ſome ſort independant of each other; moſt of the corporeal functions being carried on independant of *the will* of the ſpirit, and yet its component parts are ſubject to its inſtantaneous commands, although the ſpiritual powers can only acquire ſtrength of action in proportion to the growth, ſtrength, and maturity of the corporeal. The peculiar properties of the ſpirit, as thought, reflection, will, &c. ſeem totally independant of its corporeal companion; and yet the ſmalleſt defect, either from external or internal injuries ſuſtained by the *latter*, deprives the *former* of its powers of action, and frequently even of its rationality. The whole is a maze, incomprehenſible to all but the divine Projector and Fabricator! Let us wonder, and be ſilent!

§ 11.

Of matters infinitely great and important, we can only form our conceptions from leſſer occurrences which fall within our preſent finite powers. The hiſtories of all nations afford us frequent inſtances of ſudden, ſhort, and ſometimes dangerous inſurrections in a ſtate, which take their riſe from the very dregs of the people; but when a total revolution and ſubverſion of the ſtate and government is meditated and attempted, we ever obſerve, it takes its riſe from the machinations of the higher ranks. In this laſt ſpecies of rebellion, which is only in point, the prime inſtigators and actors are the moſt culpable, and conſequently ſubject to the ſevereſt puniſhment. Their abettors and ſupporters come next in degree of culpability, and milder puniſhment; and the multitude, who are

are drawn aſide from their allegiance rather by the errors of their judgment, than from malevolence, become objects of the ſovereign clemency.

§ 12.

We are given to underſtand by the ſcriptures before cited, as well as our own, that when the Supreme Being created the angelic bands, he conſtituted them of different tribes; ſome held highly exalted ranks; others, by due gradations, more ſubordinate ones. But although ſome of the higher tribes might poſſeſs ſuperior intellectual powers and abilities, yet all were equally endowed as *free agents*. They were not then, as in their preſent ſtate of exiſtence, fettered and chained to corrupt and mortal forms, and their intellectual powers thereby clogged and weakened; therefore *none* had a plea of exemption from puniſhment in a greater or leſſer degree, when their rebellion and apoſtacy was ſubdued.

§ 13.

WE are now to ſuppoſe (according with our hypotheſis) that the offending ſpirits are undergoing the various puniſhments allotted for them by their juſt Judge. And here it is eſſential to obſerve, that in the conſtruction of the *human form*, there ſeems to be a pre-eminence annexed, when compared with the reſt of the animal creation; although they are all indiſcriminately ſubjected, as to corporeal functions, to the ſame mortifying ſtate. There muſt have been ſome important cauſe for this pre-eminence. The offending ſpirit, under this form, ſeems to poſſeſs almoſt the full exertion of thoſe intellectual powers with which it was originally endowed;—not ſo reſpecting the reſt of the intelligent forms, which we crudely diſtinguiſh by the title of the *Brute Creation*. Between whom, and the human form, a line of limitation ſeems to be drawn by the Deity, touching the exertion of their intelligent powers, which in the brutes

ſeem

ſeem at leaſt under reſtraint, but under the human form more at large; conſequently, the former are under a ſeverer ſtate of reprobation, and lower degradation. But as to the *extent* of their intellect, as to thought, reflection, &c. we are totally ignorant; but that they are ſo endowed, is a truth beyond diſpute, although poſſibly in an inferior degree; and thoſe who attribute and confine morals to the human kind, and deny them to the brutes, only prove themſelves ſuperficial obſervers of their actions. That they are compounded beings, *ſpiritual* and *corporeal*, like ourſelves, none can deny, but thoſe who wilfully ſhut the eyes of their underſtanding. That they have ſinned, as well as we, in a pre-exiſtent ſtate, is equally certain; that they have an equal right (poſſibly a ſuperior) to expect a remiſſion of their ſins, and ſalvation at the conſummation of all things, who will be hardy enough to diſpute? David ſays, " Lord, Thou ſhalt *ſave*

" both

" both man and *beast*." What Paul of Tarsus (commonly although erroneously stiled *Saint* Paul) says more cogently on the subject, we shall reserve for a subsequent section. That man should shew *no mercy* to the brutes, is not wonderful; he exhibits none to his own species.

§ 14.

REASON (by analogy) has already suggested, that the apostate spirits were not equally guilty; consequently, in the eye of a just and merciful Judge, they could not *all* be doomed to equal punishment; therefore, it may be rationally supposed, that the Deity, in constructing the mortal forms they were destined to animate, had a retrospect to the degree of guilt of each individual, or rather of each distinct *Tribe*. Thus, we may very consistently conceive, that the most atrocious *leaders* and abettors of the celestial defection, were doomed to animate the most ferocious forms,

forms, as man, lions, tygers, bears, wolves, and every other ſpecies, known and ſhunned as *beaſts of prey*, and marked for their cruel, oppreſſive, tyrannic, and malicious diſpoſitions, either in the terreſtrial, aerial, or aquatic regions of this globe: and in like wiſe, the leſſer delinquents to animate the leſs offenſive, which do not come under the denomination of animals of prey, as uſually underſtood; ſuch as the bulk of the hoofed and horned tribes, &c.: and the leaſt offending of the apoſtate ſpirits, to animate thoſe forms which *appear to us* the moſt innocent and inoffenſive of the various animals which occupy the air, the earth, and the waters; as the greateſt part of the feathered kind, &c. And it may be alſo ſuppoſed, that it is this laſt tribe of ſpirits which occaſionally animate thoſe human forms who paſs inoffenſively through the walks of life, without any mark of conſpicuity annexed to their character; as is, in general, the ſtate of the *females* of that ſpecies.

§ 15.

On a critical review and compariſon of the animal tribes, we ſhall find, that their felicity and infelicity are in wide extremes; and all are ſo conſtituted (as an immediate and perpetual puniſhment) to be inimical to, and in the general, the natural prey and food of each other, the human kind not excepted. Man and brute's greateſt enemy *is man*; in every age, and on every portion of this habitable globe, marked with univerſal havock and deſtruction to the animal forms, the works of nature, art, and labour, we find that man has been the wicked *prime agent*. Pre-eminent as he is, in the extent and exerciſe of his intellectual powers, he is equally ſo in every ſpecies of vice; in violation of his ſuperior reaſon and rank in the ſcale of intelligent beings.

§ 16.

It is moſt conſiſtent with reaſon and probability to ſuppoſe and believe, that the Deity, in the creation of the material forms, deſtined, by a kind of *ſympathetic movement* incomprehenſible to us, that the delinquent ſpirits ſhould naturally, as we may ſay, and ſpontaneouſly, enter thoſe forms which beſt ſuited their various degrees of guilt, and diſpoſitions, without any immediate *new* effort of the Creator; and that this union ſhould *in ſuch wiſe* be perpetual, on every re-entry of the ſpirit (on the diſſolution of its former priſon) into the ſame ſpecies, or into any other form analagous thereto. For inſtance; a malignant ſpirit, being obliged to take its flight from the expiring tyger, would from that ſympathy, forcibly impreſſed upon it by the Deity, enter and animate another form of the ſame kind, or that of the lion, the wolf, &c. and thus of all the other tribes of the delinquent angels, through the various animal creation.

§ 17.

ALTHOUGH it appears difficult to comprehend how the ſufferings of the corporeal part of our compoſition ſhould affect *pure ſpirit*, yet the fact is certain; but as the diſeaſes and pains of the *body* have their riſe, in the general, from the intemperance and folly of the will in the *ſpirit*, it is but equity it ſhould participate in the higher degree, as the greater and indeed the only tranſgreſſor; by a perverſion of reaſon, in the injurious uſes and taſks it impoſes on the corporeal functions ſubjected to its dominion: hereby entailing a puniſhment all its own, and not inflicted by its Creator. Other inevitable conſequences attend this perverſion of reaſon and the will; namely, the premature diſſolution of the body, and more frequent tranſmigration of the ſpirit: but of this more hereafter.

§ 18. THE

§ 18.

THE ſpirits of angel, man, and brute, being identically, and ſpecifically, *one and the ſame free agents*, are conſequently *accountable*; otherwiſe, if not free agents, they cannot be accountable beings;—they are mere nothings, air and matter, machines only, actuated and moved like puppets, by ſome unſeen power behind the curtain; a concluſion, which a conſcious principle within us forbids our harbouring.——Thus far premiſed, we will cloſe this our introductory diſquiſition on the nature, and relative ſtate of man and brute, by a ſhort recapitulation of the preceding matters. We have traced the cauſes which excited the Deity to the creation of intelligent beings: He formed them with powers to obey certain injunctions he had ſubjected them to, in the obedience of which he conſtituted their prime felicity; but this obedience was *optional* and *free*. Ambition and pride urged a portion of them to ſpurn at ſub-

 ordinate

ordinate happineſs, and they rebelled; juſtice doomed them to certain puniſhments, for *a certain ſpace of time*; contrition ſucceeded, clemency and mercy interfered, and the SUPREME formed a material univerſe for their ſojourn, during *the ſpace* of their baniſhment. He framed this ſtupendous work on a GENERAL PROVIDENTIAL plan, for the preſervation and duration of the whole, on the invariable operation of cauſes to effects, and implanted in the material mortal forms, deſtined for their cloſer impriſonment, a certain principle, which ſhould excite each individual to renew and propagate its ſpecies in perpetuity. And by the laws of diſtributive juſtice, ſo conſtituted them, that every virtuous ſentiment and action, and every vice and deviation from moral rectitude, ſhould carry its *reward* and *puniſhment* along with it, (our *own feelings* confirm this truth, as do, we have no doubt, the feelings of all others) as natural conſequences of cauſes and effects;

either

either immediately, in their *present form* of existence, or relatively, in a *future*; without any interference of the Deity, until the expiration of the *term* immutably decreed for the banishment, punishment, and probation of the delinquents.——Here we beg leave to refer our readers to the perusal of our three parts of "Interesting Historical Events of Indostan," printed for Mr. Becket, London; and to our short Essay, entitled "Primitive Religion elucidated," printed for Mr. Bull, at Bath; as they will support and illustrate our present undertaking.

§ 19.

It is a well-founded principle in philosophy, that *supernatural* powers should never be called in to our aid, when *natural* prove sufficient for the purposes under consideration. When this maxim, and the various premises traced in the foregoing sections, are duly weighed in the scale of common sense

and

and reaſon, how prepoſterous, unworthy, and derogatory, muſt it not appear to the dignity and majeſty of the Divine Supreme Being, to hear his particular and partial providence or interference wantonly annexed to, and proſtituted to the moſt trifling, and degrading occurrences which paſs daily and hourly amongſt us? And is it not amazing in the extreme, that mankind, from the earlieſt records which our crude and limited chronological ideas of the duration of the world have furniſhed us with, ſhould have foſtered, broached, and propagated, ſuch a groſs, preſumptive, and ignominious conception of their GOD! Let us try if we cannot account for this ſeemingly unaccountable general infatuation.

§ 20.

IT is no leſs aſtoniſhing, that a portion of mankind, in all ages and in all nations, ſhould have acquired the addreſs to ſubjugate and hoodwink the reaſon and faculties of the reſt of their fellow creatures, in what is called ſpirituals, as well as in temporals, under the various denominations of *Prophets*, *Prieſts*, &c. Strange as this appears, the *fact* is univerſally notorious.

§ 21.

The fallen ſpirits animating this tribe, (ſtiled by themſelves *the men of God*) we may with the higheſt certainty conclude, were the very *prime projectors*, *leaders*, and moſt active *abettors*, of the revolt in heaven; and failing in their attempt againſt their God and Creator, but ſtill influenced by the ſame principles, namely, an inſatiable thirſt for power and dominion, they meditated how they ſhould ſubject their fellow-rebels to their ſway and government here below; which, taking the advantage of their original ſuperior faculties and art, they were eaſily enabled to accompliſh in the following manner.

§ 22. Their

§ 22.

THEIR firſt political movement was the aſſuming an external ſanctity of manners, and to eſtabliſh a prepoſſeſſion on the ſurrounding ſpirits of ſomething *ſacred* being annexed to their perſons and characters; and to heighten and improve this *reverence*, they pretended frequent and familiar intercourſe with the DEITY, and that at their interceſſion, he would grant every petition *they* preferred to him; and finally they inculcated the principle we are combating, that GOD, by his peculiar and partial providence, perpetually interfered in the tranſactions of individuals, and that *their* daily interpoſition on the behalf of ſinners, was eſſentially neceſſary, to ſoften and deprecate his wrath and vengeance. Thus by ſlow, but ſure degrees, they reached the ſummit of their views, and got under their abſolute controul, not only the conſciences, but the perſons and property of the bulk of the people, and retain that dominion

dominion until this hour, over ninety-nine hundred parts of this habitable globe. And thus only can be accounted for this universal infatuation, operating on the contrition, fears, and apprehensions of the multitude, upon their *recent expulsion*, and banishment from the celestial regions, by the crafty insinuations of this *malignant Tribe*; who, by the *impious tenet* of the constant interference of the Deity in the transactions of mankind, precluded the first gift of their Creator, *free agency*; thereby making their GOD the author of, or conniver at, all evil. But they stopped not here; for ambition, power, and avarice know no bounds: in process of time, they impiously assumed the prerogatives and attributes of the DEITY; they made themselves to be worshipped and adored! they possessed themselves of temporalities and principalities, and trod on the necks of kings; and by the subordinate agency of their brethren, they sowed dissention religious and civil throughout every land where they obtained a footing. Upon a retrospective view of the history of all nations, we shall find them, either openly or covertly, the active promoters of persecutions, blood and slaughter, rebellions, and murders.

§ 23.

To this erroneous principle of a *particular* and *partial* providence, are also ascribed the rise, revolutions, downfall, and extinction of mighty empires, kingdoms, and states; although the annals of the world obviously shew they all can be accounted for without the interference of supernatural powers, by the result of common and natural *causes* and *effects*. Had the Senators of Rome been annually elected, as the Consuls were, and the destructive distinction of Patrician and Plebeian never taken place, probably Rome had not fallen. Had there been a legal successor of mature age to Alexander's dominion, or

had

had he even *appointed* a ſucceſſor, probably the Macedonian empire had ſubſiſted to this day. And, in a late moſt mortifying inſtance;—had an adequate coercive force been immediately exerted by the acting powers; had theſe powers not been miſled by falſe intelligence and advice; had *all* of thoſe entruſted with the executive part, done their duty, as ſome individuals did, with exemplary loyalty, and military prowess; and had not rebellion abroad been foſtered, aided, comforted, and abetted, by traitors at home; this nation had not ſuſtained an irreparable loſs. And thus of every ſtate, that has, does, or may exiſt; there has ever been, and probably ever will be, ſome fundamental error in its riſe, or *political* progreſs, which by the natural operations of cauſes and effects, inſures its final deſtruction, without any neceſſity of intruding ſupernatural interference.

§ 24. THE

§ 24.

THE peculiar and partial *providence of* GOD, is alſo abſurdly uſhered in to account for the moſt trifling incidents of life, as before remarked. It is proſtituted in every brothel; and ſeminary of vice; it is hackneyed throughout every trumpery novel; and *implied*, (nay, is the *ſine qua non*, the fundamental,) in the worſhip of every community; for ſhould this derogatory principle be exploded, the influence of the tribe who firſt eſtabliſhed it, would ceaſe, as well as the temporal emoluments ariſing therefrom. They regarded not in what degree they depreciated the dignity of their GOD, in cultivating this tenet; although conſcious that thereby they introduced him on every occaſion, counteracting his own work of *free agency*, and totally annihilating the very idea or poſſibility of *preſent* or *future* rewards and puniſhments.

§ 25.

We may rationally conceive and conclude, that the Supreme Being, wrapped in the divine contemplation of his own beatitude, and in the adoration and fidelity of the remaining angelic bands, looks down, with an eye of benevolence and commiseration, on those who fell from their allegiance. He sees the perversion of those intellectual powers with which he had originally endowed them, as a *passive spectator only*. He sees with pity, that in place of exerting *those powers* in the laudable pursuits, and enjoyment, of such temporary local blessings as he had benevolently left within their reach, and in cherishing and protecting each other, under whatsoever *external form* they were inclosed. He observes *them* mutually exerted for the diametrically opposite purposes of deceit, perfidy, oppression, and cruelty, one to the other; or devoted to useless, vain, and presumptive researches, in pursuit of what are vaguely

termed

termed knowledge, arts, and ſciences; or devoted to every ſpecies of frivolous wanton luxury and diſſipation, which ſenſuality can dictate; or perpetually exerciſing their talents in lawleſs violence, rapine, murder, drunkenneſs, and ſuicide. He marks the multitude, ſtill under the infatuation and influence of thoſe malignant ſpirits who firſt drew them from their original purity. He notes thoſe ſpirits, inſtigating and ſtirring up the ambition, pride, and avarice of kings, to the deſtruction of each other, with millions of their miſerable ſubjects; placing their God, or his *deſtroying angel*, at the head of their armies.

§ 26.

On the foregoing ſurvey of the *real ſtate* of the inhabitants of this globe, can we poſſibly conceive, that they are the objects of a juſt God's providential care and conſtant interpoſition, with leſs than a degree of blaſphemy?—Surely, no.

§ 27.

IT will be objected, that if this be really and truly the caſe, and that the DEITY does not interfere, otherwiſe than by his general laws and providence in the government of the world and its inhabitants the *inutility* of all religion and religious worſhip is neceſſarily *implied.* Not ſo, abſolutely, as our ſubſequent ſections ſhall evidence; but the trade, the traffick, the external pomp, parade, and ceremonials of all religious worſhip exiſting, certainly are; and muſt undergo a ſevere arraignment and proſecution. Let us *ſeriouſly* (if poſſible) take a review of all the ſyſtems of religious worſhip in the known world, from times almoſt immemorial to the preſent day; let us comment on their various contradictory, idolatrous, puerile, non-eſſential, ridiculous, legerdemain tricks and ceremonies, and unceaſing ſchiſms, annexed to them all without exception; and then aſk ourſelves whether we can conceive, conſiſtently with reaſon, that ſuch a motley worſhip can be pleaſing to God, which appears juſtly ridiculous in the eyes of man?

§ 28.

We will next take a view of the daily farce and mockery of religion and religious worſhip. Two neighbouring ſtates proclaim a diabolical war againſt each other, founded on ambition, pride, avarice, punctilio, or other pretences; in the courſe of which, deſtruction dire falls on their reſpective countries, their people, and a large portion of their fellows of the brute creation; and famine and peſtilence not an uncommon conſequence! The religious worſhip eſtabliſhed in each of the kingdoms of theſe belligerent powers, ſupplicates the Deity to ſanction, aſſiſt, and ſupport their infernal operations, and *Te Deum* on each ſide is ſung for their various ſucceſſes and triumphs in the *glorious* and *pious* thirſt and purſuit of blood and deſolation.

desolation. Can the peculiar providence of a benevolent God, be possibly conceived to act or interfere in such scenes of horror? He can neither *permit* or *connive* at evil; nor can it be imagined, without the highest impiety, that he interposes or permits that medley of atrocious sin and wickedness, which are hourly perpetrated in every capital city of the world; or that he interferes, guides, or influences, the destructive operations of war, excited by malevolent motives, and carried on with unremitting infernal fury on both sides. Is not then the mockery of worship just above alluded to, most justly arraigned?

§ 29.

We will now take a view of religious worship in different lights, where the supposed interfering providence of the Deity, by the most unaccountable infatuation of the rational faculties, is invoked to change, suspend, or stop the course of his general laws, to gratify the preſent intereſted cravings, diſcontent, and impatience, of a few greedy individuals. If an uncommon drought happen, he is + worried to ſend rain upon the earth; if the contrary fall out, he is invoked to give ſunſhine and fair weather; not adverting, that his general laws of diſtribution, in either caſe, are impartial and immutable. Two ſhips in purſuit of that baneful ſcience *commerce*, (which ever has been, is, and ever will be, the perpetual ſource of miſchievous contentions between nations) are rounding a promontory, in oppoſite courſes, the one requiring a north wind, the other a ſouth; we will ſuppoſe them to be at ſuch a diſtance as to be out of ſight of each other, with various unfavourable currents of air, and foul tempeſtuous weather; in both caſes, the DEITY is + worried by prayer to comply with their contrary petitions, for fair winds and weather; and if they fail, the images of their tutelar ſaints or idols reſpectively, are brought

in proper word loose upon

out

out by their *priests*, carried round in ridiculous procession, and invoked to mediate with God for a fair wind, or to quell the storm; and if they are esteemed tardy or idle in their application, they are scourged and ducked, to bring them into more compliant order. Now, is it possible to conceive, that in nature there can be any thing so perfectly ludicrous and farcical, as these incontestible *facts* present us with? Infatuated and misguided spirits! what follies and extravagancies will ye not be guilty of, under the continued influence of your prime malignant leaders? Can you believe, that your God will listen to, or interpose his divine providence to your unworthy incongruous supplications? or that he will stretch forth his arm to save and extricate you from perils and dangers which you have avariciously and wantonly brought upon yourselves?—No, your cool and uninfluenced reason will not justify such absurd and offensive conceptions of your God.

§ 30.

THE foregoing three ſpecimens of the farce and mockery of religious worſhip ſhall ſuffice, although we could produce a multitude more, equally derogatory to the juſt ideas we ought to entertain of the Supreme Being; but a ſtricter ſcrutiny into the *principles* and *dogmas* of all religions remains for diſcuſſion, which has been only ſlightly glanced at in our 27th ſection. He muſt be little verſed in eccleſiaſtical ſtory, who can for a moment doubt that the *ſyſtems* of all, without exception, fall juſtly under the laſh of *reaſon* and *real piety*. They are unworthy GOD, and ourſelves; they are all a ſtrange compound of unintelligible ſenſeleſs jargon, and incomprehenſible contradictory *myſteries*, illuſtrated and decorated with equally ſenſeleſs forms and non-eſſential ceremonies. If they had any thing really pure, and worthy the DEITY, in their birth and original inſtitutes, they are ſo mutilated, corrupted, and adulterated,

adulterated, their authors would not know them again, but ſpurn at and deteſt the ſpurious iſſue. We have ſtudiouſly and candidly traced them all to the fountain head, and have no doubt remaining but that they were originally founded on *pure ethicks*, framed, promulged, and propagated, by ſome of the leaſt offending *contrite ſpirits*, on their firſt taking poſſeſſion of this planet; and iſſued as *laws* and principles that would inſure their local felicity, and prove at the ſame time pleaſing to their merciful Creator. On this pure baſis, the malignant ſpirits, in proceſs of time, raiſed a ſuperſtructure ſubverſive of reaſon, and all things ſacred; they formed ſyſtems of *religion* and *worſhip*, correſponding with the plan we traced in our 22d ſection, inſuring to themſelves *importance* and *emoluments:* they held forth their God as a ſtalking-horſe, a decoy, to enſnare the minds of the unwary multitude in the nets of ſuperſtitious credulity,—they ſucceeded, and eſtabliſhed

bliſhed their power and influence over them. The modern ſyſtems of BRAMAH, MOSES, and CHRIST, were founded on the unity and ſupremacy of the Godhead, and *ethicks*; the firſt and laſt on a more exalted, ſublime, and refined ſcale, and they have all ſuffered the ſame mutilated and corrupted fate, by the ſame miſchievous malignant ſpirits. We call thoſe ſyſtems *modern*, ſetting at nought the waking-dreams of the wiſeſt of the wiſe, reſpecting the *real* duration of this globe, and planetary univerſe. The Mahommedan ſyſtem, although a mere compilation of extravagant, incoherent rhapſodies, yet by ſtealing from Moſes the pious tenet of the *unity of the Godhead*, he eſtabliſhed to himſelf the character of a divine Prophet and meſſenger from GOD, and his Koran was rapidly embraced by nineteen twentieths of the delinquent ſpirits of this earth, forming the greateſt empires exiſting.

§ 31.

WE have occaſionally arraigned the purſuits of mankind, in ſearch of *knowledge*, *arts*, and *ſciences*. We ſhall not in this expoſition tritely quote Solomon as our oracle, although by long experience he found them all futile and *vain:* our aim is, to prevent the miſapplication of time, expence, and talents, which might be employed to better purpoſes. Whether we ſucceed or not, it muſt be allowed that our intentions and attempt are laudable; in order to which, we will, with deference, analiſe a few of the moſt capital, and begin with the moſt ancient—Aſtronomy.

§ 32.

THE Chaldeans, according to our limited records and ideas, are ſuppoſed to have been the firſt aſtronomers. Their obſervations on the annual revolutions and motions of the viſible luminaries, as far as they conduced to inſtruct them in the return of times and ſeaſons for the cultivation of the earth, was natural and proper; it was a ſimple and uſeſul knowledge, which exiſts to this day in all countries, upon the ſame artleſs principles, by a race totally ignorant of aſtronomy *as a ſcience*. Of what *real* uſe or importance is it to mankind *in general*, to know whether the ſun moves round the planets, or the planets round the ſun? or whether his place in the Zodiac is in Leo, Libra, Taurus, &c.? of what ſignification is it to know whether our globe ſtands ſtill, or has a daily rotation from Weſt to Eaſt? The ſolution of theſe, and many other phœnomena, may be ingenious; but the wiſely ignorant (if the expreſſion may be allowed) in theſe abſtruſe and unprofitable reſearches are as happy, and probably more ſo, than the *few*, whom vanity has prompted to ſignalize themſelves as beings of a ſuperior rank. The unlearned, the multitude, are ſatisfied with obſerving the uſeful obvious *effects*, without diving for profound and hidden

cauſes,

causes, which lie out of their depth. They see, with placid minds and gratitude, the sun, moon, and stars, rise and set, and the alternate vicissitudes of light and darkness, day and night; and adore the Being who first created them, and gave motion to the whole:—not so the Chaldeans; for as they are supposed to be the first astronomers, they are also deemed to have been the first idolaters, by the instigation of the malignant spirits, who from this astronomical root branched out the science of judicial astrology, which infamously taught, that from certain conjunctions, aspects, and combinations of the stars and planets, the inevitable fate of every individual was distinctly marked; and that they themselves only were their infallible interpreters. The Priesthood, in early days, arrogated to themselves this species of deceit; and to this day, in the East, continue to avail themselves of it, with considerable emoluments; but they have dropped it in the more enlightened West, and a few needy wretches amongſt the laity have taken it up, under the title of fortune-tellers, who never fail of dupes in the lower ranks of the people. On the whole, it muſt appear incongruous to reaſon and common ſenſe to imagine, that the DEITY can be pleaſed with theſe vague purſuits, and attempts to penetrate his impenetrable laws of creation. The theories of this ſcience have ever been ideal, at variance, and contradictory, the one exploding the other; and ſo it will be ad infinitum; therefore, the utmoſt that can be ſaid of the celebrated labours of its profeſſors, is, that they have exhibited a profuſion of deep erudition and exalted genius to very little purpoſe, reſpecting the well-being or felicity of the *general*; and that the ſuperior talents of Pythagoras, Ptolomy, Copernicus, Tycho Brachæ, Gallileo, Des Cartes, Newton, and the reſt of the ſtar-gazers, have been totally miſapplied: and it is our opinion, that the admired line of Pope,

"And ſhow a NEWTON as we ſhow an ape,"

(alluding to the ſuppoſed wonder of the celeſtial beings on his arrival amongſt them) which is commonly underſtood to be a high compliment paid to that great man, (for great he certainly was) rather carries with it a concealed ſatire.

§ 32.

We will next conſider the art of Navigation, to which aſtronomy and geography have been aiders and abettors; and here we may ſay, that the malignant ſpirits ſeem to have reached the *ne plus ultra*, or extreme of their malicious purpoſes, for the perpetual confuſion and deſtruction of all mortal forms. It is highly improbable, that when the Deity planted the different regions of this globe with the fallen ſpirits, or intelligent beings, his deſign was, they ſhould ever have communication with each other; his placing the expanded and oc-occaſionally tempeſtuous ocean between them exhibits an inconteſtible proof to the contrary.

trary. But in this, as in every thing elſe, man has counteracted his wiſe and benevolent intentions. *This art* firſt created diſcontent in the minds of men with the lot and ſtation marked out for them by their Creator, (for touching the ſuppoſed *diſperſion* of the original inhabitants of the earth from the plains of Shinar, we look upon it, and we hope without offence, as merely ideal or allegory, as other productions of the ſame author undoubtedly are.) It excited new cravings, and imaginary wants, and as men found they had acquired the means of gratifying them, they ſet no bounds to their paſſions and appetites; in proceſs of time they invaded every quarter of the globe, and marked their way with horror and deſolation, under the covert ſpecious pretence of *extending* commerce, (a ſcience which cannot be too often execrated) accompliſhed by the ruin and murder of millions, to glut the avarice and wanton deſires of the few: wit-

neſs

neſs the conqueſt of Mexico and Peru, the bloody tranſactions and cruelties attending the Eaſtern expeditions, the circumnavigators, buccaneers, &c. &c. Surely the interfering hand of Providence cannot be ſuppoſed to have a ſhare in ſuch atrocious deeds, without a degree of blaſphemy.

§ 33.

THE art of Printing, what dire miſchiefs has it not produced? what diſſentions, civil and religious, moral and divine, has it not excited in the boſoms of contending mortals? what ſavage cruelties, bloodſhed, and murders, has it not been the cauſe of? what deſtruction to general peace, and tranquility publick and private, has it not been the parent of? It has ever been the ſucceſsful weapon of diſcord, in the hands of the malignant ſpirits, to ſtimulate to lawleſs ambition, ſedition, and rebellion; it has propagated and diſſeminated controverſial feuds and principles, derogatory

tory to the being and majeſty of God, and his divine attributes; it has dared to publiſh and circulate productions, whoſe tendency could only vitiate and deſtroy the morals and virtues of mankind; and for one *grain* of good this art has produced, it has ſown a pound of evil, which has ſprung up an hundred-fold. The productions of Romance, Fiction, Novels, Poetry, and Muſic, although poſſibly the leaſt exceptionable of all the dependant tribes on this art of printing, have nevertheleſs, improper and dangerous tendencies; they are all calculated chiefly to amuſe, and lull to ſtupor the imaginations of thoughtleſs beings; to tickle the ears, and feed the ſenſes of the voluptuous and extravagant; and lead aſtray the minds, particularly of the youth of both ſexes, from more uſeful and eſſential applications. The tenets of a few moral writers diſſeminated by this art, may have produced ſome partial good, but the immoralities ſcattered and circulated by the ſame

channel,

channel, have overballanced them tenfold. "Some writers exalt human nature to a de-"gree of excellence they have no claim or "pretenſions to; others debaſe it to a degree "too humiliating;" although, by the *general bent*, the latter ſeems to have demonſtration on their ſide. The philoſophic diſquiſitions propagated by this art, have only ſerved to miſlead and puzzle the multitude, and confound the underſtandings of all. Some "Philoſophers have ſpiritualized matter; "others have materialized ſpirit;" ſome have affixed a preciſe birth to the duration of the world, others make it eternal. In ſhort, and to cloſe this ſection, there are no abſurdities or extravagancies that could be ſtarted in the imagination, that this art has not been the nurſing mother to, to the utter confuſion of all right and wrong; but in a land of freedom, the liberty, or rather the licentiouſneſs, of the preſs is not to be reſtrained, although every evil is attendant on it. Here it may be ſaid, with a ſneer, "You, notwithſtanding, avail yourſelf of this art, to obtrude your crudities on the public." To this we ſhall only anſwer, that we wiſh every author took up the pen from the ſame benevolent motives, the art would then not be a ſubject for cenſure.

§ 34.

POLITICS, or the arts of legerdemain, fineſſe, circumvention, deceit, and fraud. Surely it can be no offence to ſay, that the interfering hand of Providence has no ſhare in this dirty buſineſs.

§ 35.

TACTICS, or the art of war and murder. [We have juſt peruſed the particular detail of the attack and defence in a late ſiege.] O ETERNAL ONE! is it poſſible thou canſt pardon the belief and imputation, that thy Divine Providence interferes, or has any concern,

concern, in the works of this infernal and diabolical art?—If the principles of this art, taken in all its parts and direful consequences, is not a proof that the DEITY is only a *passive spectator* of the transactions of mankind, we give up the point; but will still flatter ourselves, that on mature reflection every thinking being will conclude as we do.

§ 36.

THE art of Painting in all its branches. Although we are most sensible that in this section we shall draw upon ourselves the censure of crowned heads, academical societies, and the opulent; and the scorn and resentment of the combined groupe of connoisseurs and virtuosi, yet we shall not hesitate to pronounce the general rage for this science an irrational, unprofitable, and mischievous pursuit, both in its professors and admirers; an art conceived by indolence, brought forth by vanity, nursed by affectation, and supported by pride, oſtentation, and prodigality;—whereby immenſe treaſure is ſunk, to the injury of the ſtate, in uſeleſs lumber and dead ſtock, whilſt the poor and indigent are ſtarving at our gates. But our *ipſe dixit* ſhall not here ſuffice. The ſupport of ethicks, or morals, ought to be the prime view and effort of every artiſt, let his profeſſion be what it will; let us try how far the profeſſors and productions of this art have tended to propagate them. The ſubjects ſelected by the moſt celebrated artiſts have been generally taken from Pagan theology and mythology, which cannot conduce to that great end; others have choſen ſubjects from the religions of later times, as the Annunciation, Conception, &c. which at the ſame time that they excited a veneration for the art, it perpetuated the ſuperſtitions, follies, and abſurdities it records. Our pen would bluſh to enumerate the indecent works of ſome of theſe celebrated artiſts, whoſe names ſhould

be

be ſunk in oblivion, as well as their productions, which act more powerfully and dangerouſly on the imagination than any immoralities conveyed by the preſs. But we will analiſe this ſcience a little farther, and begin with hiſtorical paintings, as firſt in dignity: This repreſents remarkable events and tranſactions, exhibits battles, ſieges, triumphal entries of heroes, tyrants, and conquerors; aims to perpetuate incidents, and the memory of a race of beings, which have been a peſt to ſociety, a diſgrace to the human form and intellect, and the bane of all moral rectitude. Not ſo was employed the immortal HOGARTH, whoſe unequalled genius and pencil laboured to impreſs, on the minds of the vicious, a moral lecture in perpetuity. Landſcape and Paſtoral Painting is a vague, futile, inadequate attempt to copy and repreſent the beauties of nature, which falls infinitely ſhort of the original, although executed by the moſt maſterly hand; therefore to what *real* uſe is this labour beſtowed? a waſte of time and talents to cover a wall, when at the ſame time a man may look out of his window, and enjoy the ſubject in much higher and tranſcendant perfection. Portrait painting can have had no motives for its ſource and ſupport, but a family pride, vanity, and oſtentation. Reaſon will whiſper, that if the object repreſented be impreſſed on the heart, it is totally uſeleſs; if it be not, the preſervation of the ſemblance is truly farcical: again, if a departed object was dear to us, and the remembrance of it a drawback on our internal peace, prudence will whiſper it cannot be too ſoon forgot. What has been advanced on painting, is equally applicable to Statuary, ancient and modern. The rage for both, amongſt the pretended Cognoſcenti, has been carried to a moſt ridiculous extreme. To cloſe this ſection, we will readily allow, that prodigies of genius in theſe arts have graced the world in all ages; the more is the pity;

as in general, this claſs of men are polite, inoffenſive in their manners, and of good morals, endowed with penetration, acute obſervation, and ſagacious: with theſe talents and qualities, they would be ornaments to, and more uſeful members of the community, if their genius had a more active bias for the *real* benefit of their fellow-creatures.

§ 37.

ARCHITECTURE, as far as it conduced to comfort, convenience, and ſhelter from the inclemency of ſeaſons, was neceſſary; but all beyond, in the eye of reaſon, is excited by vanity, profuſion, and luxury; and is futile and unſtable in the enjoyment, as every day's experience and obſervation fully evince. The great erect palaces they ſeldom viſit, and uphold them only for a ſhow, to be gaped at by the caſual paſſenger; the diſeaſe is infectious, it pervades the *little great*, and lower ranks of the people; and we may venture to aſſert,

that

that more diſtreſs, difficulties, and ruin, have been brought on families and individuals, from this miſchievous ſcience and rage of building, than from any other mad propenſity that ever took poſſeſſion of the brain of man.

§ 38.

We have yet left a large field to explore, but will ſtop here, having, to all unprejudiced minds, proved the inutility, fallacy, and miſchievous tendency, of thoſe ſciences already handled, reſpecting their imaginary *real benefit* and happineſs to mankind. The *utile dulce* of ſcience point out themſelves; as moral, natural, and experimental philoſophy, agriculture, mechanics, &c. wherein the viſible tendency promotes and ſecures the general good:

" One moral, or a mere well-natured deed,
" Does all deſert in *ſciences* exceed."

The viciſſitudes of events have been manifeſt throughout all ages; our confined records have

have ſhewn, that one part of this globe has been enlightened, whilſt another has been obſcured in blindneſs, ignorance, and barbariſm; and again, theſe enlightened, and the others ſunk in darkneſs. Nor is it at all improbable, but that thoſe arts and diſcoveries, which are, by an unwarrantable conceit, attributed as it were to yeſterday, may have flouriſhed, and been loſt *alternately*, in different regions, for a ſpace of fifty thouſand or millions of years back, for any thing we *really* know to the contrary. Chronology is one of the leaſt excuſable reſearches that has employed the genius of indolent, ſedentary men; much ſtudy and pains have been waſted, to affix and adjuſt certain æras, epochas, dynaſties, &c. to events which are of no eſſential importance. Chronologiſts have proceeded on the vague hypotheſis of the world having been in exiſtence only about ſix thouſand years; but here they all differ in their calculations, and no wonder, as they worked in the dark,

dark, without any certain principle. The chronology of the Eaſt and Weſt differ more than their latitude and longitude. The Egyptians, if we may credit Herodotus, prove the exact periods of the ſun's riſing twice in the *weſt*, and ſetting in the *eaſt*, by the revolutions of the planetary ſyſtem; the Chineſe go beyond them; and the Gentoos ſtill exceed, and calculate the birth and duration of the material formed univerſe to a day, upon principles we have ſhewn elſewhere: and after all, to what valuable purpoſe have the labours and erudition of an USHER or a NEWTON, &c. been expended, when it matters not whether the creation of the univerſe was ſix thouſand or ſixty millions of years paſt; whether this, that, or the other incident, event, or tranſaction, fell out in this, that, or any other period of time. Equally futile have been the warm and idle conteſts between nations, touching their antiquity; but more ſuperlatively ridiculous in individuals, the

pride,

pride, vanity, and folly of *pedigree*.——But to reſume the principal objects of our diſquiſition.

§ 39.

On a retroſpective view of the riſe, progreſs, ſummit, declenſion, and fall, of all empires, kingdoms, and ſtates, that ever exiſted; according to hiſtorical annals *ſacred* and *profane*, (a diſtinction, we confeſs, we do not underſtand) we ſhall find they were *replete*, through every period of their various ſtages, with fraud, violence, oppreſſion, rapine, murder, and every other deviation from moral rectitude, which a depraved and diabolical nature can be capable of. For *particular inſtances* we need not have recourſe to remote times and regions, when we are ſupplied with ample recent matter nearer home. We will take the liberty of reciting a few, which ſingularly mark the vindictive and deſtructive genius of man. Let us caſt an indignant eye

on

on the ſlaughter and devaſtation following the various invaſions of Gaul and Britain, and the long and inveterate conteſt for the ſubjection of France by the latter; the wretched and depopulated ſtate of theſe our kingdoms, during the barbarous and bloody ſtruggle between the Houſes of York and Lancaſter; the fire and faggot reign of our firſt Mary; the miſerable ſtate of a neighbouring kingdom, during the long and ſanguine conflict between Catholic and Hugonot; the obſtinate trial for dominion on one ſide, and liberty on the other, between Spain, and the United Seven Netherland Provinces; and laſtly, the civil war in theſe kingdoms, during part, and ſubſequent to the reign of Charles I. when an univerſal delirium ſeized on all ranks of the people, and furious fanatics, ſpurred on and aided by ſubtle, concealed, political fraud, overthrew the government in church and ſtate, pretending on every occaſion, an impulſe divine, under the protection of God, and his *peculiar*

providence,

providence, as a ſanction for the moſt atrocious crimes, and in violation of every thing ſacred; as the public manifeſtos, addreſſes, declarations, &c. of thoſe diſtracted times, are amply pregnant with, and fully demonſtrate.—Theſe few recent inſtances, which exiſted as it were but yeſterday, we preſent only as ſpecimens, or ſamples, of what mankind has been almoſt from his firſt deſtination under his preſent form; as from the hiſtories of all nations, ſimilar inſtances might be produced in confirmation. This being the real ſtate of the caſe, and will be ſo we fear, in perpetuity, do we not ſtand juſtified in concluding and ſaying, that if fifty thouſand *devils*, or malignant ſpirits, had been inveſted with the guidance, interference, and government of empires, kingdoms, and ſtates, and tranſactions of man, they could not, more fully, effectually, and uniformly, have deſtroyed his peace and felicity, both here and hereafter, than man himſelf has done; and yet theſe are the works we impiouſly attribute to the interference and influence of our benevolent, juſt, and merciful God; whereas, on the contrary, it fully proves that man is abandoned, and left to the free operations of his own *benevolent* or *malevolent* will. We will cloſe this part of our diſquiſition with this final concluſion, that an immediate interfering, peculiar, and partial providence, and *free-agency*, are incompatible with each other, and cannot poſſibly exiſt in the nature of things.

§ 40.

Permit us now to revert to the inſtances of our laſt ſection, and notice the active part the Chriſtian Prieſthood took in moſt of them, if not in all; but more particularly that claſs of them ſtiled Dignatories of the Church. From authentic memoirs we find, that theſe malignant ſpirits, in place of preaching up unity, peace, and love, (the

pious

pious dictates of Him, whom they sometimes, without any meaning, called Lord and Master) proved the very firebrands who set nations in a blaze! by the power and influence they had obtained over the weak minds of Kings and people, we find them incendiaries, setting Kings against their people, and the people against their Kings; and instead of inculcating the gospel of peace, charity, and obedience to the laws, we find them blowing the trumpet of sedition and rebellion, and personally wielding the sword of blood and slaughter, at the head of armies! We find them also, by the folly of their Sovereigns, possessing themselves of the first offices in the state, and assuming a jurisdiction superior to, and independant of the laws, and arrogating a rank above the nobles of the land; and other attributes they usurped, that are more particularly specified in our 21st section, to which we beg leave to refer.

§ 41. THE

§ 41.

THE foregoing portrait of this claſs of ſpirits, although more immediately applied to the Chriſtian Clergy of the times above alluded to, yet it is alſo juſtly applicable to the ſame claſs, in every nation of the globe. It was aptly ſaid, by a very competent judge of human nature, that "Prieſts of all religions are the ſame;" their power, their influence, their conſequence, and depredations, have ever been obtained under the inſidious maſk of ſome *religious ſyſtem* or other; the mockery, the fallacy, the impiety of *all*, will be obvious to every unprejudiced mind, who ſtrictly ſcrutinizes their internal and external tenets and principles:—therefore, it is full time the maſk ſhould be taken off, and mankind releaſed from the leading-ſtrings of *ſuch* religious nurſes, and reſtored to his native freedom, which has been ſhackled for ſo many ages, in ſuperſtitious bondage. It will here, probably be aſked,

What

What does this mortal, this enthusiast, this drawcansir, mean or intend, by all this indiscriminate abuse and satire? Does he wish to sow the seeds of discord, sedition, anarchy, and confusion? Is it his view to embroil the churches and states of all nations? Does he aim at overturning all religion, and religious worship?—God forbid:—his meaning, intentions, views, wishes, and ideas, are much the reverse, as the following sections will clearly prove.

§ 42.

To guard against the censure of indiscriminate satire, we, from our heart and knowledge, freely confess and believe, that there may have existed, and do exist in all nations, *some* even amongst the dignified, as well as subalterns, of the clerical class of spirits, who have abhorred and detested the pride, ambition, and avarice of the order, and have strictly and piously adhered to the duties of their sacred function only: but alas!—

§ 43.

HITHERTO our ſtrictures have been general and univerſal: we ſhall now confine ourſelves within the Chriſtian pale. Our former labours tended to eſtabliſh the ſacred doctrine of the UNITY and SUPREMACY of the GODHEAD, which, we humbly conceive, the liturgy and worſhip of every Chriſtian Church palpably oppoſed and diſcountenanced. The incomprehenſible dogmas in favour of *Polytheiſm*, none but one of the prime malignant ſpirits animating the form of an *Athanaſius*, could poſſibly have meditated, or propagated; and yet we ſtill retain theſe incomprehenſibles in our worſhip! We wiſh not the abolition of churches, the prieſthood, or religious worſhip; our aim is, to ſee them all reduced to ſuch a ſtandard as may do honour to GOD, and be conſiſtent with reaſon, true piety, and propriety. It is true, the extenſive arm of ſacerdotal power and influence, has in theſe latter times been *ſhortened*, and rendered leſs

miſchievous;

miſchievous; but it is ſtill too long, and it behoves every Chriſtian Government to *cut it off*. Permit us to expatiate on the various miſeries, perſecutions, and cruelties, excited and perpetrated by theſe malignant leaders of the Chriſtian Church, on every oppoſer of the various *changes* they have *rung* on the pure, plain, ſimple, dictates and doctrines of CHRIST, for the ſpace of ſeventeen centuries back; the recollection pains the imagination, humanity ſtarts at the idea of the numerous maſſacres and ruin poured on the heads of ſocieties and individuals; inſomuch that a benevolent mind cannot avoid execrating the fatal diſtinction of *Catholic* and *Proteſtant*, with their miſchievous tribes of diſſenters, under every denomination. The ſubject is too ſerious and important to provoke to mirth, but philanthropy may without offence beſtow a pitying ſmile on the *early* diviſion and *later* ſubdiviſions of the Chriſtian Church and its profeſſors, into Catholic, Lutheran,

 Calviniſt,

Calvinist, Independant, Puritan, Presbyterian, Anabaptist, Quaker, Methodist, Moravian, Sandimonian, with a long *et cætera*; all harbouring bitter rancour in their hearts against each other; *each* of this motley tribe claiming infallibility from scraps taken from the same scriptures, variously interpreted, by the vain, dark, designing, self-interested, malignant spirit at their head, as the different genius of each pointed out to their enthusiastic and crafty brain, sinking the others to everlasting perdition. For our own part, we profess ourselves of no particular sect whatsoever, but an adorer of ONE GOD, in *spirit* and *truth*, and an humble follower and subscriber to the *unadulterated* precepts and doctrines of CHRIST. This short confession of our faith, we think necessary, to guard against any misconception or misrepresentation of our principles. The above just stigmas are suited to every other system of religion in the world, which is in like

manner

manner divided and ſubdivided into ſectaries, and ſtrongly marks the follies and abſurdities of mankind, under the influence and guidance of wicked and intereſted leaders.

§ 44.

When a religious ſyſtem is erected into what is called a Church, and endowed with temporalities and powers, framed to overturn ſtates and kingdoms, it then becomes a mere political inſtitution, and the beſt evidence of its divine origin is deſtroyed; and when the fallacy of ſuch a religion was attempted to be undermined by the profeſſors of *pure ethicks*, the Prieſthood, alarmed for their temporalities and powers, convened themſelves together in the devil's name, and propagated a ſyſtem (originally) of meekneſs, peace, and mercy, by the arguments of *fire* and *ſword!* Enthuſiaſts may boaſt "the "influence of an enlightened religion," and draw a juſt preference to the *original* doctrines of CHRIST, on a comparison with the Koran of Mahomet; but, alas! stubborn facts are against our hopes of a conversion to this enlightened religion in the inhabitants of the East, *as it is now professed.* It is not the *real* doctrines and precepts of CHRIST, that are now either preached or practised. And does not the annals of Christendom, for a series of centuries, exhibit such instances of superstition, persecution, cruelties and butcherings, committed under the pretended sanction, influence, and title, of this *enlightened* religion, which were never yet perpetrated by Jew, Turk, or Pagan? On what grounds or basis, then, can we expect converts to such a religion, which had neither stability or uniformity, for half a century from its first promulgation? Its divine Author, and divine precepts, may virtually be said to have been a million of times crucified, since the sacrifice of himself, in support of his doctrines; the teachers of the present *enlightened* system, ſubſcribing to articles of faith they neither comprehend or believe; and every ſect aſſuming the name and title of Chriſtians, without poſſeſſing one iota of the *genuine ſpirit* of Chriſtianity, either in ſentiment, purity of *worſhip*, or *diſcipline*.

§ 45.

The above being the preſent ſtate of the Chriſtian ſyſtem of religion, throughout all Chriſtendom, without the ſmalleſt exaggeration, no one, we think, will be hardy enough to deny, that a *general reform* is eſſentially neceſſary, and loudly called for, by the voice of *true piety* and *reaſon*. The radical cure for any evil, and *its effects*, cannot be expected, without minutely tracing it to its original *cauſe*; we have glanced at this already, but now we ſhall ſpeak more openly, and without reſerve pronounce, that all the evils with which mankind has been peſtered in all ages, ſprung from an undue *pre-eminence*, *power*, and *emoluments*, aſſumed by, and weakly granted to the *prieſthood*. According to the political modes of government in ſtates, a diſtinction of *ſuperior* and *ſubordinate* ranks and degrees became eſſentially neceſſary; but as Politicks and Religion have no relation to each other, but always move in diametrically oppoſite directions, there never could have been any neceſſity for thoſe ranks, degrees, and diſtinctions, in the government of a Church, or in the adminiſtration of the holy functions. Herein our ancient politicians committed an egregious and unpardonable blunder and ſolecifm in politicks; but they knew not the artful, encroaching, Machiavelian ſpirits they had to deal with, nor adverted to the Eaſtern proverb, which ſays, "Give ſome people a finger, and they will "ſoon take your whole hand." Touching the emoluments granted to the Prieſthood, under the ſtile of temporalities and patrimonies of the Church, the impropriety of the meaſure has been manifested in all ages, by the mischievous use that has been perpetually made of them, as all history demonstrates. A learned and pious Bishop, above a century and a half back, preached and protested against temporalities being annexed to the Church, [vide Sir Richard Baker's useful and laborious chronicles] and the opening of our 44th section are the sentiments of a dignified member of the Church, deservedly celebrated as the most acute moral writer that has graced this latter age. All temporalities annexed to Popedoms, Archbishopricks, Bishopricks, Deaneries, Prebendaries, Vicarages, Rectorships, Colleges, &c. were originally alienations from the *publick stock*, surreptitiously and fraudulently obtained, by the artful suggestions and influence the Priesthood had acquired, over the minds and consciences of their respective weak and inconsiderate rulers, and sovereigns; it is notorious, and confirmed by ecclesiastical

history,

hiſtory, that all ſeminaries, colleges, &c. for the drilling and training members for the Church, were founded and endowed at the inſtigation of the clergy, for the immediate benefit of themſelves or their ſucceſſors; it is allowed they are alſo ſeminaries for the advancement of *uſeleſs* arts and ſciences: but, query, have they not likewiſe proved ſeminaries of vice, libertiniſm, and lewdneſs, to the detriment of real morals and virtue?—Let it not therefore appear ſtrange or ſingular, or ſacrilege in us, our attempting to diveſt the Church of its temporalities; in fact, it has no *legal* right to them, nor is it the Church we diveſt; it is the over-fed poſſeſſors of them we propoſe to ſtrip, that they may revert in juſtice and equity to the original rightful owners, THE PUBLIC, and be appropriated to the relief of the preſent heavy and deplorable exigencies of the STATE. And in this, we have no doubt but we ſhall be ſupported by the votes of ninety-nine in the

hundred

hundred of the inhabitants of every Chriſtian country; and the Clergy themſelves, if they poſſeſs a grain of conſcience, *publick ſpirit*, and love for their country, (which we will not doubt) will not heſitate a moment to ſay Amen; eſpecially as it is not our intention to ſend the Prieſthood a grazing, but only to reduce them to a *reſpectable level*. The abject ſlavery and dependance of the ſubaltern Clergy has long been a reproach to the Church and Legiſlature of every Chriſtian Government; but if our plan takes place, there ſhall not be a *ragged Curate*, in his Majeſty's dominions at leaſt. As Propoſitions are now become the mode of addreſs on all occaſions, we ſhall adopt it in our next ſection.

§ 46.

Propoſition 1ſt. It is humbly propoſed, that the dignified Clergy under every denomination, be diveſted of all Rank, Precedence,

and

and Title, in the Church and State; ſaving and excepting that of *Doctor in Divinity* only, which every member of the Church ſhall indiſcriminately enjoy, on the ſame reſpectable and rational level.

Prop. 2. That a period be put to the long, miſchievous, illegal, and irreligious practice of mixing temporalities with ſpiritualities; and that all endowments of whatſoever kind, annexed to Cathedrals, Churches, Chapels, and Colleges, be ſequeſtered, *reſtored*, and appropriated to the relief of the exigencies of the ſtate, and heavy burdens of the people.

Prop. 3. That the forms of ordination, ſubſcription, and *degrees*, be totally aboliſhed as uſeleſs, and to the full as farcical as the *noli epiſcopari*; and that the KING, as ſupreme head of the Church, ſhall, by himſelf, or by delegation to his Miniſters of State, occaſionally *ordain* and *preſent* men of ſound

and

and tried morals and underſtanding, to the beſt of their knowledge and information; profound learning, and knowledge in the dead languages, being abſolutely *non-eſſentials*.

Prop. 4. That a conſiderable reduction ſhall be made in the number of churches, and every church be independant, and but one incumbent to each church; and in caſe of ſickneſs, or other inability, his place ſhall be ſupplied by the incumbent of the next adjoining pariſh, on proper notice given of the neceſſary variation of the hour for the commencement of the ſervice.

Prop. 5. That a ſtipend of five hundred pounds per annum, exempt from all taxes, office fees, and deductions whatſoever, be eſtabliſhed for every married incumbent, and three hundred for every one unmarried, in lieu of all tithes, ſurplice-fees, and other perquiſites, with a decent and commodious

parſonage

parſonage houſe, handſomely furniſhed at all points, in the vicinity of the church, the whole to be kept in repair by the government; the ſtipend to be paid from the treaſury half yearly, the firſt half year in advance upon their induction: as theſe ſtipends are more than equal to landed eſtates of eight and five hundred pounds per annum, it cannot but be deemed a *reſpectable proviſion.* Whether the exigencies of the ſtate can admit of this ample proviſion, the legiſlature alone are the competent judges.

Prop. 6. That the reduced Dignitaries, in compenſation for their loſſes in temporalities, ſhall have the preference on the new preſentations taking place, and next to them the doctors in divinity of the preſent eſtabliſhment; but as our plan only propoſes one incumbent for each church, the ſupernumeraries in orders, which overflow the land, muſt not be left to ſtarve; therefore we propoſe

pose an annuity of one hundred pounds shall be settled upon them for their lives; and as many of them will drop off annually, the state will be soon released from that burden.

Prop. 7. That the Doctors shall still retain the honorary titles, insignia, and emoluments, (if there are any) of being chaplains to his Majesty and the nobility.

§ 47.

THE foregoing equitable propositions put a stop to the oppressions and grievances of ecclesiastical courts, to the scandalous and shameful *trade* and *traffick* of religion, as practised in all Christian churches, for simoniacal pluralities, lay presentations, dispensations, &c. they set every member of the church upon an equal and respectable footing; they preclude tithing, pregnant with dire mischiefs and cruelties in our sister kingdom, and the never-ceasing source of enmity,

enmity, rancour, and contention, between the clergy and laity; and consequently raise the value of landed property throughout the realm; and are also pregnant with many other salutary consequences, to the honour of God and true religion, and the ease and essentially necessary emoluments of the state; and detrimental only to a very few individuals in comparison with the whole distressed community. We will readily admit, that the present dignitaries of our established church are as pious, learned, and respectable a body, as ever graced any age or nation; but still they possess and riot in the spoils of the Public, sedulously, fraudulently, and illegally obtained by their predecessors, and therefore cannot in conscience maintain their right to them. We are sensible, however, we shall draw on ourselves the bitter indignation and resentment of the dignified priesthood of all Christian churches, (it must be allowed they have some provocation) but we shall

shelter

ſhelter ourſelves under the conſcious integrity of our intentions for the good of the common-weal, and leave the defence of our cauſe to *their ſubalterns*.

§ 48.

THE moſt arduous part of our various ſubjects, *Religious Worſhip*, now only remains for diſcuſſion; and here we muſt lament our inadequate abilities for ſo important a matter, relying on the indulgence of our readers, and hoping that wiſer heads will correct, ſupply, and fill up, the deficiencies of our imperfect *outlines*, for we will not preſume to arrogate to ourſelves any higher merit.

§ 49.

WE cannot open our intereſting ſubject better, than with the words and ſentiments of the great Chancellor HYDE, in his celebrated ſpeech to both Houſes of Parliament, (the ſectariſts of thoſe times) immediately after the reſtoration of Charles the ſecond; "How would they (the primitive Chriſtians) "look upon our *ſharp* and *virulent* conten-"tions and debates on the *Chriſtian Religion*, "and the *bloody wars* that have proceeded "from thoſe contentions; whilſt every one "*pretended* to all the *marks* of the *true church*, "except only *that* which is inſeparable from "it, *charity* to one another." Theſe were the times, as before remarked, of general diſtraction, when fanaticiſm and enthuſiaſm rode triumphant. But what ſhall we ſay to thoſe earlier times, and to thoſe very *primitive* Chriſtians, whoſe examples were only followed on the ſame principles, by thoſe of later times. The early ſeparation of the Greek Church marked the inſtability of the ſyſtem, and the ſame virulent contentions and debates, and the ſame bloody wars, ſoon ſucceeded, ſtirred up and inſtigated by thoſe then (and now to our reproach) ſtiled *Fathers of the Church*. From them thoſe diſſentions,

and

and conſequent cruelties, have had a regular deſcent, without intermiſſion, to this hour; and enthuſiaſtic reformers have ſtarted up in every corner of the Chriſtian world, without one grain of *charity* one to another. Such have been the direful conſequences of torturing Chriſtianity into a non-eſſential *ſyſtem*, which its original tenets ſtood in no need of. The torrents of blood which have been ſhed in theſe *Chriſtian irreligious* diſſentions, have ſtamped an indelible ſtain on humanity, never to be waſhed out whilſt records and memory exiſt: it was not the cauſe of God, or *religion*, which excited theſe deplorable ſtruggles; the real objects were, *power* and *temporalities*.

§ 50.

It has been manifeſt, from the earlieſt periods ſucceeding the death of its founder, that Chriſtianity had no *ſettled* and *uniform* principles or doctrines; and that like rays diverging from a center, the farther it extended, the wider its professors differed in their tenets and sentiments; each sect asserting and arrogating to itself the *infallible marks* (as just above hinted) of the *true Church*, by proofs drawn from the same fountain, namely THE SCRIPTURES, variously interpreted, as their speculative interested views and fancies dictated, until at length they left no *precise meaning* to any parts of them: the real divine principles, and pure ethicks, they taught, were overwhelmed and frittered away, and in their stead, systems of non-essential and incomprehensible tenets, and unintelligible jargon, were instituted, and even these without any uniformity whatsoever, as the various and inconsistent liturgies of every Christian Church exhibit ample proof of. In such a perplexed state and situation, what then remains for a rational and anxious enquirer? He cannot possibly do otherwise than indignantly spurn at and

reject

reject every liturgy exiſting, as warring with reaſon, true piety, common-ſenſe, and each other; and unworthy our GOD and ourſelves. Benevolently moved for the preſent and future well-being of our fellow creatures, but more particularly for the ſtate under whoſe protection we breathe, we will humbly attempt to ſketch out ſuch a form of religious worſhip as ſhall not be liable to any of the above juſt ſtrictures and objections, and to which all unprejudiced rational beings will not heſitate to conform; but previous thereto, it becomes eſſentially neceſſary, firſt, to remove one ſtumbling-block out of our way; and ſubſequently, to analize the preſent liturgy of our eſtabliſhed Church, and try it by the laws of REASON and PROPRIETY.

§ 51.

WE are aware that ſome will urge againſt us the old hackneyed plea, that in attempting religious innovations, there is no knowing where things will ſtop; witneſs the *Reformation*, (as it is called) which was not eſtabliſhed without much bloodſhed, broiling and roaſting of fooliſh individuals, as well as attended with dangerous commotions in this ſtate, and other parts of Chriſtendom. But let it be remembered in anſwer, that the genius of men in theſe our days, are widely different from the ſuperſtitious and fanatic times above alluded to; *prieſtcraft* has loſt its power and influence, and mankind now feel their right of thinking for themſelves; therefore we venture to aſſert, that any juſt and neceſſary reformation in religious worſhip may be as eaſily and peaceably eſtabliſhed, under a ſteady adminiſtration, as any *other* neceſſary reform in the ſtate; and in this aſſertion we are ſupported by a clauſe in the 34th Article of our Religion, which ſays, " Every particular or national Church hath " authority to *ordain*, *change*, and *aboliſh*, " *ceremonies* or *rites* of the Church, ordained

" only

" only by *man*'s authority,—ſo that all things
" be done *to edifying*."

§ 52.

HAVING thus cleared our way, we will proceed in our analyſis. This is not the firſt time we have ſubmitted to the public our ſtrictures on the Liturgy of our Church, as incompatible with the true Chriſtian religion, as dictated by its founder. In this we have not been ſingular: many learned and pious members of the Church have meditated a *partial reform*, utterly inadequate to any rational purpoſe: our aim is a TOTAL ONE. And firſt, the incongruity of mingling in our worſhip the hiſtory, infamous wars, and indecent events, of a race of people, deteſted and execrated by all nations, is glaringly reprehenſible; the wonder is, how it has obtained ſo long, to the diſgrace of human wiſdom and propriety! But we hope and truſt, the period is not far diſtant,

when

when neither reaſon or the chaſte ear ſhall be offended, or lulled to ſleep, by the tedious, unintereſting recital of a ſeries of tranſacti-ons, which only tend to mark the depravity of the ſpecies throughout all ages and na-tions; nor continue longer *half Jew*, and *half Chriſtian*.

§ 53.

THE next, and moſt reprehenſible parts of our Liturgy, are thoſe which pay an ado-ration to CHRIST equal to GOD himſelf, thereby palpably impeaching the UNITY and SUPREMACY of the GODHEAD; of which egregious error Mahomet took every advan-tage, in the propagation of his abſurd Koran, to the material detriment of the Chriſtian faith. We need add little more, on this dangerous and derogatory principle of wor-ſhip, than the ſentiments of CHRIST himſelf, who openly and ſtrongly oppoſed and pro-teſted againſt any ſort of adoration being offered

offered up to him; immaculate as he truly was, his humility would not even ſuffer himſelf to be ſtiled *good*, as none but God, he ſays, was really ſo. Therefore, let us in ſincerity of heart, pay every *due* reverence to his name, life, ſufferings, and precepts; but let us not longer devote that worſhip and adoration to him, which he pathetically tells us is *only* due to *his* God, and *our* God, to *his* Father, and *our* Father, which is in heaven: But regardleſs of this, we in our worſhip *bow* at the name of Jesus, whilſt the name of our God paſſes unheeded.

§ 54.

No leſs fatal and dangerous are thoſe parts of our Liturgy, which countenance the efficacy of *Mediation* and *Atonement*. Howſoever theſe doctrines may have been ſophiſtically maintained and ſupported by intereſted prieſtcraft, it is *morally* impoſſible to conceive, that the mediations, ſufferings, and death of

any

any one individual, can atone with a juſt God for the crimes of millions, or even of *one ſinner*; and we heſitate not to aver, that theſe tenets have operated more to the deſtruction of all morals, than any other cauſe whatſoever. We daily ſee that thieves, houſe-breakers, and murderers, whoſe lives have been one continued ſeries of rapine and violence, depend at the gallows (by the fallacious encouragement of the prieſthood) on the remiſſion of their ſins, by the death and mediation of Christ, upon a momentary repentance, when the laws of the land put it out of their power to tranſgreſs any further, at leaſt under that mortal form. We have no doubt, but that in ſupport of the doctrine of *atonement*, the 28th verſe of the 20th chapter of Matthew will be urged againſt us; it runs thus, "Even as the Son "of *Man* came not to be miniſtered unto, "but to miniſter, and to give his life as a "*ranſom for many*." Taking the meaning of this

this verſe in its utmoſt latitude, it cannot poſſibly imply more, than by giving his life in ſupport of his miniſtry and doctrines, *many* would be ſaved or *ranſomed* from their ſins, by religiouſly adhering to his precepts. In this ſenſe, his death becomes a virtual ranſom or atonement; not that it could, in a literal ſenſe, be an atonement for the ſins of the whole world, as frequently expreſſed in the Liturgy; for if ſo, he would be counteracting his own precept of the neceſſity of *repentance*. Yet on this weak foundation, the fatal and dangerous conſtruction was erected by ſome ſectariſts, that *faith* was alone ſufficient for ſalvation, without *good works*; thereby precluding the neceſſity of repentance and amendment, and making moral and free agency only an ideal conception of the imagination. In the verſe above cited, and the context in the preceding ones, CHRIST ſeems merely to be giving a leſſon of humility to his Apoſtles, in the miniſtration of his doctrines;

trines; he instituted no Popes, Patriarchs, Archbishops, Bishops, Deans, &c. he tells them "the Princes of the Gentiles exercised "*dominion*," and adds, "*but it shall not be so* "*among you*;" plainly intimating and prohibiting their exercising any dominion or superiority over one another; and yet we have seen Popes, while they humbly stiled themselves *servus servorum*, exercising universal dominion and tyranny over Churches, Dignitaries, Kings, Princes, and Potentates; the same spirit of dominion and oppression actuating every rank and degree of the sacerdotal order throughout all Christendom; those *above*, oppressing, spurning, kicking, and trampling on those *below*, in violation of their Master's injunction, that there neither should be *high* or *low* among them.

§ 55.

Divine worship should be as purely spiritual as possible, unmixed with all indifferent

temporal

temporal concerns, or political heterogeneous matters, which reaſon tells us cannot merit the notice or attention of GOD; here it is obvious, we allude to thoſe parts of our Liturgy which conſecrates and commemorates the Fifth of November, the Thirtieth of January, and the Twenty-ninth of May: they have no rational tendency, they ſerve only to keep old wounds of the conſtitution open, and contention alive. We may with propriety execrate and condemn a daring and infamous attempt of inflamed inſtigated zealots againſt the ſtate, detected and prevented by a mere *natural* incident of regard and friendſhip, in one of the conſpirators; the conſecration of the day might have anſwered a political purpoſe at the time, but now we may juſtly pronounce this part of our Liturgy a proſtitution of divine worſhip. Equally reprehenſible is the conſecration and commemoration of the death of an unhappy Monarch, who being miſled by precedent, and weak

weak and evil councils, oppreſſed the liberty and property of the ſubject, and in the end, by a fatal combination of cauſes and effects, brought his head to the block. But moſt reprehenſible is the conſecrating the reſtoration of a man, whoſe morals and politicks, in every point of view, rendered him unworthy of commemoration.

§ 56.

Next in turn for diſcuſſion, comes what are termed the official ſervice of our Liturgy, in which we apprehend weak minds may gather cauſe of offence; but we fear God more than man: and firſt, of the Sacrament of the Lord's Supper, a ſubject we ſhall labour to treat with that becoming decency and reverence it calls for; although we may, with all humility, diſſent from the inſtitution, in the principle, letter, conſtruction, mode and manner of its adminiſtration. In the courſe of this ſacrament, as well as in the

Litany,

Litany, and in almoſt every other part of the Liturgy, is manifeſtly impeached the UNITY of the GODHEAD, more particularly in the recital of the Creed, where CHRIST is ſtiled "God of God, Light of Light, very God of "very God," &c. which we cannot conſider in any other wiſe, than as glaring blaſphemy againſt the Supremacy of the DEITY. The whole proceſs of this ſervice falling eminently under the ſtrictures of our 52d and 53d ſections, to which we may add the long, tedious, formal, and non-eſſential parade of unintelligible repetitions, and needleſs quotations from ſcripture and canons, the too frequent mention of the body and blood of Chriſt, the conſecration of the ſimple elements of bread and wine, to almoſt a ſpecies of idolatry, notwithſtanding the ſalvo at the end of the Communion ſervice. Here the doctrines of Mediation, Atonement, and Saviour by proxy, form the very baſis of this inſtitution, and the innocent and immaculate

late Chriſt is made the ſcape-goat for the remiſſion of ſins and ſalvation, and burdened with not only the crimes of Iſrael, which were abundantly ſufficient, but with the ſins of the whole world, millions of whoſe inhabitants never heard of his name or doctrines. It has been a moot point amongſt the learned, whether this inſtitution, ſo ſimply and *unequivocally* enjoined by Chriſt, was intended by him in perpetuity to *all*, or appropriated *ſolely* to his immediate Apoſtles and Diſciples, delegated to propagate his doctrines. Be this as it may, it is a truth incontrovertible, that it would have been happy for ſucceeding generations, had it been always underſtood in the latter ſenſe, and never erected into an official inſtitution of the Church; for then would have been ſpared the deluges of blood, wickedly ſpilt in the contentions reſpecting the principle, and various modes of its adminiſtration. We cannot cloſe this ſection, without arraigning the proſtitution of this

ſacrament

ſacrament on various occaſions; it is now made a political teſt, or qualification for offices of truſt and emolument; and by many never taken from any *other call.* It is adminiſtered to dying perſons, under the ſanction of a *death-bed repentance,* without any previous ſcrutiny into the *religion* or *morals* of the individual, whoſe life *may* have been a conſtant violation of *both.* This was not the kind of repentance preached by Chriſt; but on this *forlorn hope* does man ſin on, from day to day, with the futile and groundleſs expectation of forgiveneſs at the laſt. It is alſo adminiſtered to notorious malefactors, previous to their execution, upon a ſpecious and inadequate contrition. The vicious will eaſily ſet at nought moral rectitude, on the flattering preſumption that a ſecure paſſport will be given them at the cloſe of their iniquities, for pardon, peace, and happineſs in a future ſtate of exiſtence; although their lives have exhibited a continued ſeries of oppreſſion, fraud, rapine, and invaſion, of the rights, liberty, peace, property, and lives, of their fellow-creatures, in one ſhape or other.

§ 57.

NEXT comes under our view, the Sacrament (as it is called) of the Baptiſm of infants and grown perſons. We have, in a former eſſay, proved the total inutility of this ceremonial, as well as the impropriety of every ſpecies of *Creeds*, or profeſſions of faith; on which there have been ſuch various and miſchievous contentions. In a chriſtian country, it muſt be neceſſarily underſtood, that every perſon who enters the church *is a Chriſtian*. The teſts of Baptiſm, and *a Creed*, might *poſſibly* have been neceſſary in the very primitive times of Chriſtianity; but is now, beyond doubt, utterly uſeleſs, as the farce made of it fully proves. When we advert to the queries of the prieſt and anſwers of the ſponſors, in infant baptiſm, it is ſurely difficult

difficult to reſtrain the bluſh on the countenance of the parties. He, with ſolemn face and diction, expatiates on "the remiſſion of ſins, by *ſpiritual regeneration*, and invokes the *ſanctification* of *water*, to the *myſtical* waſhing away of ſin," &c.; and they as ſolemnly bind themſelves to perform every injunction he lays them under, although not one in a thouſand ever after trouble their heads about their wards, or the ſolemn obligations they entered into *before* God.— Chriſtian kings, tyrants, generals, heroes who have ſlain their thouſands and tens of thouſands, and every thief that has made his exit at the gallows, have *all* had their baptiſm, and their ſponſors. The inference to be drawn, we leave to the *real* Chriſtian reader.

§ 58.

Respecting the other official inſtitutes of the Liturgy, namely, The Church Catechifm, and Confirmation, Matrimony, the Vifitation of the Sick, Burial of the Dead, Churching of Women, and the Commination; after we have entered our general proteft againft them *all*, as militating againft the *unity* and *fupremacy* of the GODHEAD, and other fundamentals of pure worfhip, we purpofe to beftow a fhort fection on each.

§ 59.

THE Church Catechifm, and Confirmation, are compofed of doctrines, matters, and things, as well as terms, concerning which the wifeft heads have differed; confequently they come not within the reach of the capacity or underftanding of children, or youth; yet they are taught to repeat it like parrots, to no one prefent or future ufeful purpofe; as they hardly ever think of, or revert to it, after they arrive to the years of maturity; befides, in feveral paffages, the very idea of free-agency is deftroyed. And here

we

we cannot refrain condemning the infatuated, premature, uſeleſs, if not dangerous cuſtom of making children gallop through the Scriptures, before their underſtanding can receive any proper impreſſion from them; they are laſhed on from Geneſis to the Revelations, and ever after diſregard, or deteſt them, as objects of their early grievances.

§ 60.

THE moſt holy ſtate of Matrimony, as it is *moſt* profanely ſtiled in the Liturgy, cannot be looked upon by thoſe who enter into it in that light; for daily proofs contradict the ſuppoſition; therefore it is unworthy a place in any ſpiritual inſtitution of divine worſhip. Hourly experience proves, that it is conſidered by *all*, as a mere civil, political, temporal, convenient trading *compact*; violated at will by both, or either of the parties, under certain penalties; and totally diſſolvable by human laws; conſequently, it ſhould in future be considered and entered into as a *simple contract* only; and be divested of the mockery of prayers and benedictions; of passages bordering upon the ludicrous, and of solemn vows and covenants, which neither party in general ever think of, or regard after their departure from the altar; although their God is invoked as a witness to their mutual obligations. All wilful perjuries are severely punished by the laws, except those daily committed *in this most holy state of Matrimony!*

§ 61.

In a former essay, often alluded to, we have stated the impropriety of a formal service for the Visitation of the Sick; as particular circumstances, and situations of the individual, should be the guide of the officiating Minister.

§ 62.

THE ſervice for the Burial of the Dead, when diveſted of every part which intrenches on the unity and ſupremacy of GOD, and which implies adoration of Chriſt, may be retained, as the fineſt ſpecimen of ſentiment and language exiſting.

§ 63.

THE publick ſervice for the Churching of Women is a moſt indelicate tax upon the ſex, and gives cauſe to the congregation, and eſpecially amongſt the youth, for many indecent alluſions. If ſuch a ceremonial may be thought neceſſary by weak minds, let it be confined to the private apartment of the individual.

§ 64.

THE laſt, and moſt tremendous part of our Liturgy, is the Commination, or day of *curſing*, where the benign, good, gracious;

and merciful God, is introduced furiously denouncing *wrath* and *vengeance*, unbecoming the idea we ought to entertain of Him. We have no doubt but that most of those official institutes were originally established with a view to the fees and emoluments annexed to them; but our propositions put the Clergy on such a respectable footing as precludes such low considerations.

§ 65.

We cannot quit this part of our subject without paying *due* notice to the Thirty-nine Articles of Religion, as fundamentals of our faith and worship; they were established in the reign of Elizabeth, anno 1571, and subscribed to by the archbishops, bishops, and the whole body of the clergy, in convocation assembled. It will not, we think, be disputed, that these Articles were *ordained by man's authority only*; and therefore we may, we hope, without offence make a little

free

free with them; as they impoliticly ſigned their own condemnation in the clauſe of the 34th Article, recited verbatim in our 51ſt ſection; and ſurely it is almoſt as incomprehenſible as the Articles themſelves, that any learned and rational body of men ſhould ſubſcribe to ſuch a medley of blaſphemous, enthuſiaſtic, contradictory rhapſodies. The only apology that remains for them is, that thoſe were the times of inflamed bigotry and ſuperſtition. But it is ſtill a greater marvel that they ſhould deſcend, without variation, to theſe our more enlightened days; and ſtill remain a teſt which every one of the prieſthood muſt ſubſcribe to, if he aims at poſſeſſing any emoluments of the Church. And yet ſo prevalent and tempting is the proſpect, they, in the general, ſcruple not to ſubſcribe to Articles derogatory to the dignity of their God, and beyond the reach of human underſtanding or belief; ſacrificing every thing that is ſacred to tempora poral views and hopes of future promotion. We have heard it urged in their defence, that their reaſon and conſciences are ſhackled by a fatal neceſſity and dependance, from the mode of their training and education, which renders them unfit for any other purſuits in the ſtate; if ſo, they cannot too ſoon be re-leaſed from this *opprobrious Teſt*; the machi-nations, effuſion, and ſcum, iſſuing from the ebullitions of diſtempered brains, to the ſub-verſion of true religion, common ſenſe, and reaſon.

§ 66.

From the foregoing analyſis, it muſt be conſpicuouſly obvious to every thinking be-ing, how few parts of our eſtabliſhed Liturgy are admiſſible in a rational worſhip of the Deity. Some ſettled form of divine worſhip is eſſentially incumbent on every well-regu-lated ſtate; but then if that form be unwor-thy the Deity, it were better *they had none*;

and

and if we consign every syſtem of religious prayer and worſhip extant, to the tribunal of *true* philoſophy, reaſon, and piety; they will *all* aſſuredly ſuffer a ſentence of condemnation. Although the Deity does not interfere in the riſe or fall of empires and ſtates, nor in the thoughts or tranſactions of intelligent beings, otherwiſe than by his primæval, general, providential laws; yet we are not from thence to conclude they paſs unnoticed by Him. Inconceivable as it is, by what medium, mode, or uniſon our thoughts, imaginations, and reſolves, are inſtantaneouſly conveyed to the knowledge of God; yet the *principle* is good and eligible. The efficacy of *prayer* has been doubted by ſome learned divines, and ſeems to be diſcountenanced by Christ himſelf, except in the ſhorteſt, and moſt comprehenſive way poſſible. He ſays, "But when you pray, uſe "not vain repetitions as the Heathen do." Yet if prayer be an error, we err at leaſt on the

the right ſide by the performance, provided the conſtruction of ſuch prayer be worthy the acceptance of a benevolent God; to which purpoſe our humble efforts ſhall be exerted, in the courſe of the remaining ſections.

§ 67.

If the Communion be deemed eſſential in a new Liturgy, it muſt be ſuppoſed that every communicant comes properly prepared, with a contrite heart, in charity with all mankind, ſtedfaſtly purpoſing to ſin no more; this muſt be premiſed, or they are utterly unworthy to approach the table. Therefore, where is the neceſſity of the parade of non-eſſential preambles, exhortations, prayers, &c.? Let them all be rejected; and the following ſimple mode of adminiſtration be adopted. The Prieſt, adminiſtering the bread and wine jointly or ſeparately, ſhall ſay:—*Take this, or theſe, in*

commemoration

commemoration of the sufferings and death of Jesus Christ; who died in confirmation of his divine doctrines; the practice of which doctrines only, are adequate to the remission of sins, and indispensably necessary to our present well-being, and future salvation. And in this unequivocal wise let this Sacrament begin and end, as comprising every relative essential.

§ 68.

As weak fond mothers would think their infants devoted to perdition without the ceremonial of Baptism, (for we will not call it a Sacrament) it must in some sort be retained in the church; but here we recommend the same *abridgement* of superfluous matter as enforced in the preceding section, and the following short form: The Minister, sprinkling the infant with water, shall say: *I Baptise Thee in the Name of* God, *and receive Thee into the Pale and Bosom of the Christian Church.* Some, we doubt not, will

start

ſtart at our omitting the ſign of the Croſs; but as we look upon this part of the uſual ceremonial, as a virtual adoration, and a remaining relick of that ſuperſtitious idolatry we condemn in our neighbours, we think it ſhould be wholly rejected. Here we are tempted to a ſhort digreſſion, for which we hope the indulgence of our readers. In fact, the boaſted reformation of the proteſtants, on their defection from the Mother Church, does not exhibit any traces of *material* variation, if we except the ſuppreſſion of monaſteries, nunneries, auricular confeſſion, and the *comforts* of matrimony, reſtored to the clergy. We all know that ſeparation did not flow from any ſpiritual or religious motives, but from carnal and diametrically oppoſite cauſes; we have varied in diſcipline, whilſt we continue in profeſſion and practice the ſame reprehenſible Tenets. We have arraigned their principles of purgatory, and withholding the ſcriptures from the laity, (as

repugnant

repugnant to the word of God) without sufficient consideration. The spirit's perpetual succession to animate other mortal forms, on the dissolution of its present prison, is a *virtual purgatory*, and an immediate reward or punishment for their virtues or vices, in their preceding form of existence. The deceptive and lucrative trade made of this principle by the Priesthood of the Roman Church, is doubtless highly culpable. The shutting up the *scriptures* from the laity, was certainly a stroke of wisdom; and had that prohibition been continued by the Protestants, much mischief, folly, and madness, had been prevented. The whole Christian Priesthood have been at loggerheads perpetually about their true and various readings and significations; hence the attack on the supremacy, spiritual and temporal powers of the Head of the Roman Church, and separation from it; and *hence* arose the insane multitude of sectaries amongst the protestants

teſtants; every hot-brained zealot aiming to be a reformer, and head of a ſect; until at laſt, by this wanton and indiſcriminate uſe of the ſcriptures, we daily ſee Coblers and Taylors in our ſtreets become Popes and Pretenders, to expound them with *infallibility*.

§ 69.

Our Articles forbid the invocation of *Saints*, as repugnant to the Word of God, yet by our Creed we are taught to have faith in their communion. In the Collects, Epiſtles, and Goſpels, they are diſtinctly conſecrated and commemorated; and if this does not amount to a virtual invocation, it has no meaning; therefore, in either ſenſe, they are improper to be retained in any form of divine worſhip. Too frequent Holidays, as they are miſcalled, are injurious to the ſtate, and unintereſting to religion; and ſerve only to encourage idleneſs, debauchery, and drunk-

enneſs

enneſs in the people, which is the general mode of celebration.

§ 70.

To enumerate the remaining abſurdities of our Liturgy, and in truth the Liturgies of all other Chriſtian churches, would be an endleſs taſk; therefore the foregoing ſhall ſuffice. It remains that we endeavour to ſeparate the *gold* from the *droſs*; and from the former conſtitute and *ſubjoin* a connected, ſhort, and rational form of worſhip, worthy of our Creator and ourſelves; and to which none can have the leaſt reaſonable objection. Preparatory thereto, we ſhall premiſe, that—True religion ſhould only conſiſt of adoration to God! confeſſion of guilt, contrition, and thankſgiving, and of a conſonant form of worſhip. If we conceive that mankind is wickeder in theſe our days, than they were in antecedent times, our concluſions are not well founded: David tells us pathetically, but we think rather too ſeverely, that "The "Lord looked down from heaven upon the "children of men, to ſee if there were any "that would underſtand, or ſeek after GOD; "but they are all gone out of the way, they "are altogether become abominable, there is "none that doeth good, no, *not one*." And we have no doubt but the Davids of all the preceding ages might juſtly have ſaid the ſame: in ſuch a predicament we ſtand at preſent; we hourly and univerſally ſee the exalted rational powers of mind and body, with which we were originally and graciouſly inveſted, perverted to every evil purpoſe and purſuit: whence then that preſumptive claim, right, or expectation, which are implyed in the numerous, tireſome, vain, and contradictory addreſſes of deprecation and ſupplication, ſcattered throughout the Liturgy we arraign and condemn? and whence the flattering, unmerited, deceptive conception of the interfering, peculiar providence of the Supreme,

to

to extricate us from evils, which are ſolely the reſult of our own folly and madneſs?

§ 71.

The whole tenor of our ſections mark our predilection for the hypotheſis of our being *the very apoſtate angelic beings*, and the doctrine of the ſpirit's tranſmigration through all animated organiſed mortal forms; and we confeſs we ſee no incongruity in a firm belief of either. It is on all hands agreed, that we are placed here in a ſtate of puniſhment, degradation, and probation; if ſo, it muſt have been for ſome lapſe or crime committed in a pre-exiſtent ſtate:—a perſuaſion in the firſt leads to a retroſpect of our original dignity, and would ſtimulate to deeds that do not diſgrace it. All theſe doctrines incontrovertibly conſtituted *the Creed* of our forefathers, and are to this day the firm perſuaſion of millions, and ſome of them the wiſeſt of mankind, in various parts of the globe; and if univerſally embraced, would at leaſt have one ſalutary effect; it would work a happy change in favour of the miſerable brute creation, who are looked upon and treated as mere material machines, devoid of feeling, and of any future ſtate of exiſtence: not ſo thought the ſublime philoſopher, moraliſt, and orator, PAUL of Tarſus, [vide his Epiſtle to the Romans, in the ſervice for the fourth Sunday after Trinity;] and David ſays, "Lord, thou ſhalt ſave both man and beaſt," as before quoted. Without the aid of theſe doctrines, how ſhall we account for ſome phœnomena that daily occur? Two children born of the ſame parents, nouriſhed from the ſame breaſt, trained and educated in the ſame mode, and under the ſame inſtructors; the one ſhall prove an honour to the ſpecies, a bleſſing to the parents, and a uſeful member of ſociety; the other, a diſgrace to human nature, a curſe to the parents, and an enemy to all ſocial virtues. History affords *eminent* inſtances to our purpoſe, and demonſtrates, that there is no neceſſity for the exiſtence of any other devil, than the devil within us: the mortal form of the firſt, we may ſuppoſe is animated by a ſpirit of the leaſt offending tribes of the angelic apoſtates; the other, by one of the moſt malignant offenders. Again; the amazing unconquerable unſhaken affection, antipathy, partiality, averſion, and other feelings, that inſtantaneouſly ſtrike the ſenſorium *at the firſt ſight* of an object, can only be ſolved by the ſympathetic or electric ſtroke of a *kindred* or *adverſe* ſpirit, which animates the object viewed.

§ 72.

NOTWITHSTANDING our thoughts on muſic, as given in our 33d ſection, we have no *antipathy* to the ſound of an organ; on the contrary, we propoſe no church ſhall be without one, in proportion to the magnitude of the church, and that a conſiderable portion of our New Liturgy shall be conducted in the Cathedral stile. Nothing so much excites to devotion, lifts the soul to heaven, and impresses an awful feeling and idea of the glory of the DEITY, as the divine harmony of *sacred musick:* to this intention the art should chiefly be appropriated, in place of employing it to the purposes of idle amusement and highly expensive dissipation of *time* and *fortune*, as is now, and has been for some years back, the universal infatuated impulse of all ranks of the people.

§ 73.

WE cannot help applauding what may be stiled the *magnificence* of divine worship, respecting the decorations of our churches, and the vestments of our priesthood, in which particulars we certainly are very remiss; and as we shew a predilection in the general, for the fashions of our neighbours, why should we not imitate them in these essentials? We call them eſſentials, becauſe theſe exteriors impreſs the multitude with an awe and reverence, not only for the place of worſhip, but for thoſe deſtined for the ſervice of it. We wiſh to ſee the *diſmal black* baniſhed, the officiating veſtments of the Doctors in Divinity ſumptuouſly ornamented, and their common habit *purple*, diſtinguiſhed as the *uniform* of the Church; which colour ſhould be prohibited all other ranks.

A NEW

A

NEW LITURGY;

OR,

FORM

OF

COMMON PRAYER.

A NEW LITURGY.

WHEN the wicked man turneth away from his wickedneſs, and doeth that which is lawful and right, he ſhall ſave his ſoul.

If we ſay that we have no ſin, we deceive ourſelves, and the truth is not in us. But if we confeſs our ſins, and truly repent, God is juſt, and will forgive them.

DEARLY beloved brethren, the Scripture moveth us in ſundry places to acknowledge and confeſs our manifold ſins; and that we ſhould not diſſemble nor cloke them before the face of Almighty God our heavenly Father, but confeſs them with an humble,

humble, lowly, penitent, and obedient heart; to the end that we might obtain forgiveneſs of the ſame. And although we ought at all times humbly to acknowledge our ſins before God, yet ought we moſt chiefly ſo to do, when we aſſemble and meet together, to render thanks for the great mercies and benefits we have received at his hands. Therefore I pray and beſeech you, to accompany me with a pure heart and humble voice, ſaying,

ALMIGHTY and moſt merciful Father, We have erred and ſtrayed from thy ways like loſt ſheep. We have followed too much the devices and deſires of our own hearts. We have offended againſt thy holy laws. We have left undone thoſe things which we ought to have done; and done thoſe things which we ought not to have done; and there is no goodneſs in us. O Lord, have mercy upon us miſerable offenders.

ders. Spare thou them, O God, who confeſs their faults; and *reſtore* thou them that are penitent; according to the divine doctrines promulged, preached, and practiſed, by the moſt perfect of thy created beings, Jeſus Chriſt; by which we ſhall hereafter live a godly, righteous, and ſober life, to the glory of thy holy name. *Amen.*

HYMN.

REND your hearts and not your garments,
And turn unto the LORD your God:
For he is gracious and merciful,
Slow to anger, and of great kindneſs.

ALMIGHTY God, the Father and Creator of all that exiſt, who deſireth not the death of a ſinner, but rather that he may turn from his wickedneſs and live; and pardoneth and abſolveth all them that truly repent, and unfeignedly believe his holy

Goſpel, through the pure doctrines of Jeſus Chriſt: ſpare us good Lord.

HYMN.

TO the Lord our God
Belongeth mercy and forgiveneſs,
Although we have *rebelled* againſt him;
Neither have we obeyed
The voice of the LORD our God,
To walk in his laws, which he ſet before us.

OUR Father which art in heaven, Hallowed be thy name; Thy kingdom come; Thy will be done on earth, as it is in heaven: Give us this day our daily bread; And forgive us our treſpaſſes, as we forgive them that treſpaſs againſt us; And leave us not in temptation; but deliver us from evil. *Amen.*

HYMN.

WE acknowledge our tranſgreſſions,
And our ſins are ever before us.
The ſacrifices of God are a broken ſpirit:
A broken and contrite heart,
O God, thou wilt not deſpiſe.
Repent ye, for the kingdom of heaven is at hand.

O GOD, merciful Father, that deſpiſeth not the ſighing of a contrite heart, nor the deſires of ſuch as be ſorrowful; mercifully accept the prayers we make before Thee, in all our troubles and adverſities, whenſoever they oppreſs us.

HYMN.

O COME, let us ſing unto the Lord,
And heartily rejoice in the hope of our ſalvation.
Let us come before his preſence with thankſgiving,

And

And ſhew ourſelves glad in him with
psalms.
For the LORD is a great God,
And a great King above all Gods:
In his hands are all the corners of the earth,
And the ſtrength of the hills is his alſo.
The ſea is his, and he made it,
And his hands prepared the dry land.
O come, let us worſhip, and fall down,
And kneel before the LORD our Maker;
For he is the LORD our God.

WE humbly beſeech thee, O Father, mercifully to look upon our infirmities; and for the glory of thy name, turn from us thoſe evils which we moſt juſtly have deſerved, by our *original apoſtacy* from thy holy laws. Our whole truſt and confidence, O God, is in thy mercy, whilſt we adhere to and follow the pious dictates and pure doctrines of Jeſus Chriſt. *Amen.*

HYMN.

HYMN.

WE praiſe Thee, O God:
We acknowledge Thee to be the Lord.
All the earth doth worſhip Thee:
The Father everlaſting.
To Thee all angels cry aloud:
The heavens, and all the powers therein.
To Thee Cherubim and Seraphim
Continually do cry,
Holy, holy, holy!
Lord God of Sabaoth;
Heaven and earth are full
Of the Majeſty of thy Glory.

O ETERNAL ONE, with unbounded gratitude, love, and adoration, for all thy mercies, all thy bleſſings, which Thou haſt graciouſly beſtowed on us, and all mankind; We bleſs Thee for our creation and preſervation; and with contrite hearts preſume to look up to Thee, our God and

Creator, our ultimate hope and comfort, for pardon, not only of our great *original* transſgreſſion, but for our accumulated ſins in this our preſent ſtate of *puniſhment* and *probation*. And we bleſs thy lenient hand, and unmerited clemency, moſt humbly depending that in thy good time, thou wilt mercifully deliver us from theſe corrupt and mortal forms, and *finally reſtore* us to thy divine preſence, from which, for our diſobedience, we were moſt juſtly baniſhed. This we hope for, O Lord, through the efficacy of the pure doctrines of Jeſus Chriſt, to whoſe name and memory be all due praiſe and honour paid, for evermore. *Amen.*

HYMN.

O BE joyful in the Lord, all ye lands:
Serve the Lord with gladneſs,
And come before his preſence
With a ſong.

Be

Be ye ſure that the Lord he is God;
  It is he that hath made us,
And not we ourſelves: we are his people,
  And the ſheep of his paſture.

O go your way into his gates
  With thankſgiving,
And into his courts with praiſe:
  Be thankful unto Him,
And ſpeak good of his name.

For the Lord he is gracious,
  His mercy is everlaſting:
And his truth endureth
  From generation to generation.

Miniſter.

GOD ſpake theſe words, and ſaid, Thou ſhalt have none other gods but me.

People. Lord, have mercy upon us, and accept our endeavours to keep this thy law.

Minister. Thou ſhalt not make to thyſelf any graven image, nor the likeneſs of any thing that is in the heavens above, nor in the earth beneath, nor in the water under the earth. Thou ſhalt not bow down before them, nor worſhip them, for I am the Lord thy God.

People. Lord, have mercy upon us, and accept our endeavours to keep this thy law.

Minister. Thou ſhalt not take the name of the Lord thy God in vain: for the Lord will not hold him guiltleſs that taketh his name in vain.

People. Lord, have mercy upon us, and accept our endeavours to keep this thy law.

Minister. Remember that thou keep holy the ſabbath-day. Six days ſhalt thou labour; but the ſeventh is the ſabbath of the Lord

thy

thy God: in it thou ſhalt do no manner of work, thou, and thy ſon, and thy daughter, thy man-ſervant, and thy maid-ſervant, thy cattle, and the ſtranger that is within thy gates.

People. Lord, have mercy upon us,and accept our endeavours to keep this thy law.

Miniſter. Honour thy father and mother, that thy days may be long in the land which the Lord thy God giveth thee.

People. Lord, have mercy upon us, and accept our endeavours to keep this thy law.

Miniſter. Thou ſhalt do no murder.

People. Lord, have mercy upon us, and accept our endeavours to keep this thy law.

Miniſter. Thou ſhalt not commit adultery.

People. Lord, have mercy upon us, and accept our endeavours to keep this thy law.

Minifter. Thou fhalt not fteal.

People. Lord, have mercy upon us, and accept our endeavours to keep this thy law.

Minifter. Thou fhalt not bear falfe witnefs againft thy neighbour.

People. Lord, have mercy upon us, and accept our endeavours to keep this thy law.

Minifter. Thou fhalt not covet thy neighbour's houfe, thou fhalt not covet thy neighbour's wife, nor his fervant, nor his maid, nor his ox, nor his afs, nor any thing that is his.

People. Lord, have mercy upon us, and accept our endeavours to keep this thy law.

ANTHEM.

ANTHEM.

BLESSED is the man that hath not walked
In the counſel of the ungodly;
But his delight is in the law of the Lord,
And in his law will he exerciſe himſelf
day and night.

And he ſhall be like a tree planted
By the water-ſide;
That will bring forth his fruit
In due ſeaſon.
His leaf ſhall not wither, and look
Whatſoever he doeth
It ſhall proſper.

As for the ungodly, it is not ſo with them;
But they are like the chaff,
Which the wind ſcattereth away
From the face of the earth.

Therefore the ungodly ſhall not be able
To ſtand in the judgment:

Neither the sinners in the congregation
Of the righteous.

But the Lord knoweth
The way of the righteous:
And the way of the ungodly
Shall perish.

The SERMON.

ANTHEM.

THE heavens declare the glory of God,
And the firmament sheweth his handy work.

One day telleth another:
And one night certifieth another.

There is neither speech nor language,
But their voices are heard amongst them.
Their sound is gone out into all lands,
And their words into the ends of the world.

In them hath he ſet a tabernacle for the ſun:
Which cometh forth as a bridegroom from his chamber,
And rejoiceth as a giant to run his courſe.

CHORUS.

THE Lord deſcended from above,
And bow'd the heavens ſo high;
And underneath his feet he caſt
The darkneſs of the ſky.

On cherubs and on cherubims
Full royally he rode:
And on the wings of mighty winds
Came flying all abroad.

BENEDICTION.

MAY ye all, by a due exertion of the intellectual powers with which your Creator has ſufficiently endowed you, promote your own well-being: and may you, by by adoring your God, loving your neighbour as yourſelves, and practiſing the dictates of Jeſus Chriſt, inſure to yourſelves contentment here, and in due time, joy and happineſs everlaſting. *Amen.*

§ 74.

HAVING closed our short, connected, rational, and unexceptionable Liturgy, or Form of Common Prayer, we beg leave to enumerate a few of its obvious happy effects, respecting the Minister and Congregation. *He* will here find his yoke easy, and his burthen light; he will be relieved from a *drudgery*, which exhausts his corporeal and mental powers, with the additional mortification of being constrained to utter language, sentiments, and principles, which he knows must be offensive to his GOD, and are contrary to his own conscience, faith, and judgment. But now he enters the desk, and ascends the pulpit, with a heartfelt joy and gladness, and a spirit of true devotion, he never before experienced. The Congregation will not have their appetite for devotion glutted and palled with long tiresome repetitions of the same unintelligible matter, to the confounding and bewildering of their ſenſes and underſtanding, nor be lulled to repoſe by tedious hiſtorical recitals, totally unintereſting; but their ſouls will be kept alert, and alive to the adoration of *one object only*, and awake to the pious inſtructions and doctrines, which will be conveyed to them from the pulpit.

§ 75.

THE hiſtories of all times are pregnant with glaring and flagrant proofs of the pulpit being too often appropriated to moſt unworthy purpoſes; therefore our ſyſtem is incomplete, unleſs we point out certain ſtrictures, to which the preacher ſhall be ſubject. Although our ſhort ſyſtem of divine worſhip interdicts almoſt the whole reading ſervice of the preſent eſtabliſhed Liturgy, yet we by no means prohibit the preacher the apt and proper uſe of them, occaſionally, provided he ſteers clear of, and does not touch, and founder himſelf upon, thoſe

dangerous

dangerous rocks marked in our *chart*, upon which true religion has hitherto ſplit, and been caſt away. The duties of the pulpit, incumbent on the Miniſter, are, to enforce gratitude, love, and adoration of the DEITY, and a becoming ſubmiſſion to his laws and decrees; to inculcate the neceſſity of pure ethicks, as that heart, wherein moral recti-tude is not an inhabitant, can know no true peace; to impreſs the neceſſary love of our neighbour, an *eſſential* repentance, and a due veneration *only* for the name and re-membrance of Chriſt, without any alluſions of his equality with his GOD, or to his miraculous conception; *properties* which he himſelf never aſſumed or glanced at. He muſt ſtudiouſly abſtain from meddling with the myſterious nothings in the Revelations, and from drawing concluſions from prophe-cies; prophecy implying a *fatality* in events, which deſtroys the principle of free-agency. He muſt be ſtrictly prohibited the diſcuſſion

of

of political subjects, to which the pulpit has too frequently been prostituted. For all salutary purposes, he can never be at a loss for proper texts, taken from the divine doctrines of Christ, the elaborate morals of Paul of Tarsus, and from numerous select parts of the Old and New Testament; carefully avoiding all abstruse non-essential points of speculative divinity and theology.

CONCLUSION.

Rational and candid Reader!

OUR labours are at an end; they have not been excited by any affectation of singularity, but have been spurred on by a pure spirit of benevolence to all intelligent beings. Singularity of sentiment cannot have been our motive, becauſe we have no doubt but that millions think as we do, although, from a certain cautious apathy, and indolence of diſpoſition, or the active ſcenes of life they are engaged in, they have not had courage, or leiſure, to preſent their ſentiments to the public. But for our own part, as our thread of life is ſpun fine, and probably will ſoon break; we wiſh (before our lot takes place for animating ſome other mortal form) to leave a legacy to our fellow-creatures, worthy their acceptance; and which, if properly prized, will aſſuredly conduce to their preſent and future felicity.

Our view has been, to defend the honour and dignity of our CREATOR, from a fatal *miſconception:* to expoſe the fallacy, inadequacy, and inconſiſtency, of all Chriſtian religious worſhip: to extricate mankind from the ſuperſtitious, abject ſlavery they have for ages groaned under, to a *tribe* of their own

ſpecies:

ſpecies: to arraign the folly and inutility of what are called arts and ſciences, and to ſtimulate the genius, ſtudy, and abilities of men, to more worthy and uſeful purſuits: to relieve the preſent and future exigencies of the ſtate, and heavy burdens of the people, by a moſt equitable and neceſſary meaſure: and finally, to inſtitute a form of worſhip more worthy of our GOD, and of ourſelves.

Howſoever we may have failed in the executive part of our various views, our intentions are laudable, and claim the candour and indulgence of the public tribunal: what ſucceſs our labours will be attended with, time alone can develope. To what *degree* the ſtate may benefit, by the ſequeſtration of all Church and College endowments, we muſt ſubmit to better calculators, for we confeſs ourſelves little verſed in the minutiæ of the ſubject; but we ſhould imagine the amount

muſt

muſt be important. Be this as it may, *three certain*, moſt eſſential, advantages would reſult from the meaſure:—There would be an end to the complex ſyſtem of eccleſiaſtical laws and government, which conſtitute a diſtinct, independant, and improper juriſdiction in the ſtate;—our GOD would be more rationally and better ſerved;—and our Clergy more honourably ſupported, and conſequently more revered and reſpected.

FINIS.

www.ingramcontent.com/pod-product-compliance
Lightning Source LLC
Chambersburg PA
CBHW021815190326
41518CB00007B/611
* 9 7 8 3 3 3 7 0 2 5 4 3 4 *